GOATS
& their Profitable Management

GOATS
& their Profitable Management

— 9th Edition —

by
Henry Stephen Holmes Pegler

BIOTECH

2017

Biotech Books
Delhi - 110 035

2005, Henry Stephen Holmes Pegler (b. 1848–)
Original Title: The Book of the Goat: Containing full Particulars of the Various Breeds of Goats and their Profitable Management

ISBN 81-7622-148-1

Published by : **BIOTECH BOOKS**
1123/74, Tri Nagar,
DELHI – 110 035
Phone: 27382765,
e-mail: biotechbooks@yahoo.co.in

Showroom : 4762-63/23, Ansari Road, Darya Ganj,
NEW DELHI - 110 002
Phone: 23245578, 23244987

Printed at : **Chawla Offset Printers**
New Delhi – 110 052

PRINTED IN INDIA

PREFACE TO THE PRESENT EDITION

IT is now a great many years since the first edition of this book was published. Since its introduction there has been a steady increase in the popularity of goat-keeping, especially during and immediately after the war years when a scarcity in the nation's milk supply presented an opportunity—readily accepted by goat-keepers—for goats' milk to occupy a prominent position in the dairy world.

Another factor which has contributed to the present-day interest in goat-keeping has been the recognition of the objects of the British Goat Society on the part of the Ministry of Agriculture. This was first shown by co-operation in the movement to introduce, by importation, fresh blood into the herds of these Islands. Again, by financial assistance in bringing about a new Stud Goat Scheme under which grants have been made annually since 1924 towards the administrative expenses and the paying of premiums to owners of approved stud goats stationed in various districts throughout the Kingdom.

A further feature which should be mentioned in this connection is the establishing of affiliated societies, working on the lines of and in connection with the parent institution, but whose activities are mainly restricted to the respective counties from which their names are derived. The work of these local bodies has been directed towards enlisting the interest of cottagers and people of small means to the advantage of goat-keeping and stimulating competition by offering prizes in goat classes, often in connection with village flower-shows.

In the way of literature and from the point of view of propaganda, the publication by the British Goat Society, in addition to its Monthly Journal, of a Year Book has proved a most successful innovation. The numerous contributions from the pens of experts on goat management, interspersed by technical articles from known scientific authorities, have gone far to dispel the ignorance and prejudice which has ever been a more or less serious obstacle to progress in enlightening the British public on the benefits of goats' milk and the culture of the goat.

H. S. HOLMES PEGI

CONTENTS

LIST OF PLATES

THE BOOK OF THE GOAT

CHAPTER I

INTRODUCTION

THE Goat has been appropriately termed "The Poor Man's Cow," and certainly no better designation could be found to express the position of this useful creature amongst our domestic animals. In this respect it may be placed in the same category as the pig, for it converts waste vegetables and other refuse matter into milk as the latter transforms such substances into meat. To a certain extent indeed the goat holds even a higher position than the pig, inasmuch as it provides the food of mankind from his earliest infancy, furnishing those very elements of nutrition so necessary to build up the foundation of a robust and healthy constitution, the greatest blessing of life. All doctors are agreed that milk, and plenty of it, should be the principal diet of children in the early stages of their existence, if a perfect development of the frame is to be secured. Milk, however, is not generally available in England as it should be. In the towns it is usually obtainable without much difficulty, though its quality is often questioned, but in villages and the outlying cottages the article is rarely tasted except by those who can afford to keep their own cow. This may sound absurd to some people who have never personally inquired into the matter, but it is nevertheless a fact. The demand for milk in the towns and cities, and the facilities afforded by railways for its conveyance thereto from the most remote parts, entirely drain the rural districts, where the farmers, who contract to send away every drop they can obtain, object to supplying small quantities retail. It often happens, indeed, that milk is most scarce in the very neighbourhoods whence the largest supplies are procured; consequently the cottagers' children, after they are weaned, rarely taste it otherwise than skimmed. In the character of the poor man's cow, therefore, the goat is peculiarly adapted to supply this great want, in doing which it may be regarded almost as a national benefactor, for by improving the health and physique of the rural, if not the general, population it is contributing to the well-being of the community.

The position that the goat deserves to occupy amongst the people of this country, and that which it actually holds, are, however, two different things. In spite of its useful qualities it is still far from being recognised to the extent that it merits. This is not the case in Ireland nor in most parts of the Continent, where it is highly prized for the virtues of its milk. In the Emerald Isle large numbers are bred annually, and although they lead a rough life, they contribute materially both by their milk and flesh to the welfare of the Irish peasant.

As regards goat-keeping in England it may be urged, and with truth, that this country is not so well adapted to the animal as are many parts of the Continent, on account of the climate and the large area of land under cultivation. The naturally roving disposition and well-known mischievous propensity of the goat are its great drawbacks, placing it at a disadvantage in a closely-cultivated country like ours as compared with other stock, it being impossible to pasture them in herds like sheep and cattle unless in charge of a goatherd, on account of the damage they do to trees and hedges.

Goats are accordingly only met with to any extent in mountainous regions and tracts of land adjoining these. This accounts for the large number in proportion to the population in countries like Norway, Switzerland, Spain, and Portugal.

CHAPTER II

ORIGIN AND EARLY HISTORY OF THE GOAT

THE goat is classified by naturalists as belonging to the genus *Capra*, order *Ungulata*, sub-order *Artiodactyla* (goats, pigs, sheep, oxen, etc.). There is but little doubt that the numerous varieties of the domestic goat are all descended from the Psáang, or *Capra ægagrus*, a species common all through Asia Minor, Persia, and extending even into Scinde. The fossil remains of some species of goat which have been found in the newer Pliocene deposits, as, for instance, at Walton in Essex, seem to corroborate this opinion. "The jaw and teeth," writes Professor Owen, to whom the fossil remains were entrusted, "agreed in size and configuration with the same parts in the common goat, and also in the sheep ; and the highly interesting question which of these had existed contemporaneously with the mammoth and the rhinoceros was satisfactorily determined by the cranial fragment. In its shape and size, and especially in the character of the cores of the horns which were two inches in length, subcompressed, pointed, and directed upwards, with a slight bend outwards and backwards, it closely agreed with the common goat (*Capra hircus*), and with the short-horned female of the wild goat (*Capra ægagrus*). In the sheep, the greatest diameter of the horn is across the longitudinal axis of the head ; in the goat it runs almost parallel with it—a character well shown in the present fossil. Whether the *Capra ægagrus* or the *Capra Ibex* should be regarded as the stock of the domesticated goat of Europe has long been a question among naturalists ; the weighty argument which may be drawn from the character of the wild species, which was contemporary with the *Bos primigenius* and *Bos longifrons* in England, is shown by the present fossil to be in favour of

Capra ægagrus." (Owen's "Brit. Foss. Mam.," etc., p. 490.) That the *Capra ægagrus* of the mountains of Asia occasionally crossed with some allied species, as with the *C. falconeri* of India, is not improbable, and the idea is held by M. Brandt and other authorities.

The goat appears to have been domesticated from very early days, as is evidenced by the remains which have been found; though these are generally in a very fragmentary condition. According to Professor Rütimeyer, the ancient inhabitants of the lake-dwellings in Switzerland seem to have used the goat more frequently than the sheep; the goat of these ancient lakemen, which apparently differed in no respect from the race now common in Switzerland, was probably brought hither by some people migrating northwards from the Mediterranean countries. The late Professor Rolleston, who had opportunities for examining the very rich collections of animal bones from various lake-dwellings, found that "the goat is richly and unambiguously represented in the stone-age lake-dwellings, and more abundantly indeed than the sheep in the early stone-age lake-dwellings of Moosseedorf." It seems, however, to have lost this numerical preponderance towards the end of the stone period, and to have become comparatively scarce in the bronze age. M. Kinberg, Stockholm "(Internat. Congrès Anth.," p. 831), tells us that in Sweden "*La Chèvre* (*Capra hircus*) *parait avoir été primitivement plus rare que le mouton. Elle est rare du moins dans les sépultures de l'âge de la pierre de la Vestergtölande.*"[1]

These facts are entirely in keeping with the suspicions hinted at above, and with the view that our domestic animals, though coming in the ultimate resort from the East, did not reach the regions north of the Alps directly from the East, but only by passing northwards from the Greek and Italian peninsulas. For the goat, as has been repeatedly observed from the time of Aristotle ("Hist. An." ix. 4) down to the present, bears cold less well than the sheep, whilst every traveller in sunburnt barren countries may observe with gratitude and wonder what copious supplies of milk are obtained from it,

[1] "The goat (*Capra hircus*) appears to have been in its primitive state scarcer than the sheep. It is rare at least in the sepultures of the stone-age of West Gothland."

often off but limited areas in these surroundings, and from but shrubs and weeds.

The goat possesses certain advantages over the sheep as a domestic animal in "a barren and dry land where no water is," but in a palustrine or lacustrine district it possesses none. And I submit, therefore, that the abundance of it in the Swiss lake-dwellings can be reasonably explained by supposing that it was carried thither by a people or tribe migrating northwards from the Mediterranean countries. ("British Barrows," Greenwell and Rolleston. Appendix, pp. 740-1, Note.)

The wild goat of the East was doubtless in very early ages domesticated and made subservient to the wants of man; and in course of time, as the inhabitants migrated westerly, taking their live stock with them, the goat was introduced into Europe and other countries. In the East, however, was its original home, and especially in many districts of Persia, where it is now found in a great range of climate; it is in Persia, therefore, that we ought perhaps to have the earliest records of the goat, but unfortunately nothing historically is known of the early periods of Persia, and it was not till the wars of Alexander and his successors that the Greeks formed any real conception of the position and character of the land from which their ancient and formidable enemies took their name. The word *Pásang* is the name of the male of the wild goat, and signifies in Persian "the rock-footed"; the female is called *Boz*, or *Bozpásang*, though the term *Boz* is also applied to both sexes of the common domestic goat. ("Zoology of Persia," Blandford, p. 90.)

A traveller writes: "The stately march of the he-goat before the herd, and his haughty bearing, as well as the dauntless stare with which he scrutinises a stranger, are well known by all familiar with the East; and the he-goat is still commonly applied by the Arabs as a simile for dignity of manner and bearing." The Jews sacrificed goats as offerings to Jehovah; their milk was an important item of food; goats' hair was used for weaving into cloth; the inflated skins for bottles and swimming-bladders. The flesh of the kid was held in high esteem, and it was much more frequently used as food than that of lambs, large numbers of which were reared for the sake of the wool; calves were generally regarded as too expensive a luxury, except on some festive occasion. Hence one can see the full force of the prodigal's complaint in Christ's parable—"Thou never gavest me (even) a kid, that I might make merry with my friends: But as soon as this thy son was come . . . thou hast killed for him the fatted *calf*" (Luke xv, 29, 30).

The goat of Palestine (*Capra mambrica*, Linn.) is a well-marked variety of the common *C. ægagrus*; it has thick pendent ears, often a foot long. The prophet Amos probably alludes to this long-eared goat when he speaks of a shepherd taking "out of the mouth of the lion two legs or a piece of an ear." The Syrian goat, which is larger than our English goat, has long black hair and thick recurved horns. It should be mentioned that in the North of Palestine the variety known as the Mohair goat (*Capra angorensis*, Linn.), which has long silky hair, is occasionally bred.

The Assyrians made use of the goat for the same purposes as their relatives, the Hebrews: figures of the domestic goat occur on the monuments; it has high horns either curving backwards or nearly erect; in the former case they divaricate, in the latter they are nearly parallel. The cuneiform names of the he-goat, *a-tu-du* and *tsap-pa-ru*, answer to the Hebrew *'attûd* and the Chaldean *tsâphîr* already mentioned, and have the same meaning of a he-goat. The skins were employed for various purposes; among others as swimming-bladders or buoys. After removing the head and legs the skin was prepared, perhaps steeped in tannin, and filled with air. There may be seen on the monuments now in the British Museum representations of Assyrian fishermen, sitting in the water, riding cross-legged on these inflated skins; and on the bas-reliefs which show Assurnatsir-pal's campaigns (*circa* 884 B.C.) figures of fugitives swimming to the fortress are seen, each one using an inflated goatskin as a buoy. The Assyrians frequently make mention of wild-goats—*ya-e-li* (comp. the Hebrew)—probably denoting thereby the Asiatic ibex, noticed above, which is found in the hilly parts of Assyria and the adjacent countries. The *Capra ægagrus* also occurs wild, and doubtless was known to and hunted by the Assyrian kings and nobles.

The ancient Egyptians seem to have possessed large flocks of goats and sheep from very early times; for if Manetho, in the dynasties of Egyptian monarchs as quoted by Africanus and Eusebius, be correct, the goat was well known during the second dynasty.

The ancient Greeks and Romans paid great attention to the rearing of goats. Anyone at all familiar with classical authors will remember how frequently these useful animals are mentioned, especially in the Greek pastoral poems.

The fifth idyll of Theocritus gives us interesting information about goats and goatherds ; to it I must refer the reader. The Latin writers on agriculture ("Scriptores Rei Rusticæ"), as Varro, Columella, and others, have discoursed on goats, their management, their diseases, and the various breeds. From what they have written it appears that the Romans divided their goats into two classes : I. Those which had fine hair and sawn-off horns. II. Those with shaggy hair, whose horns were allowed to grow. A good he-goat, according to Columella, should have the following qualities :—It should show under the lower jaw two small warts (*verruculæ*) hanging from the neck, and should have a large body, thick legs, a full and short neck, flaccid and very long shiny hair.[1] The Romans used either to shear[2] or to pluck the hair from the goats ; as they did also with their sheep ; the Latin term *vellus*, "a fleece," is derived from *vello*, "I pluck," referring to this custom.

The jaw appendages or tassels referred to above have proved a puzzle to many people. A question on the subject was put to me by the late King Edward when, as Prince of Wales, His Royal Highness visited the "Bath and West" Show at St. Albans. All that I could tell at that time was that, roughly speaking, those little ornaments were supposed by authorities to be false or incipient ears! Mr. J. L. Whytehead, however, contributed in the 1926 Year Book an instructive article on

[1] It is curious to notice how the Roman writers, as Varro and Columella (practical agriculturists of their day), Pliny and others, lay stress on the importance of goats having these abnormal jaw appendages, the possession of which was supposed to indicate excellence of breed and a prolific nature ; whereas by Sanskrit writers these appendages were properly regarded as of no importance whatever. In the Hitopadésa (Introduction, 1, 26) a man destitute of virtue is "like the throat-nipple of the he-goat—his birth is useless"; the Ajagala-stana, *i.e.*, "goat throat-nipple," is an emblem of any useless person or thing.

[2] From Lucian ("Piscator," 46) it appears that the same custom prevailed among the Greeks. *Tragokourice machaira* was an instrument for shearing he-goats.

this subject which I venture to reproduce here. Mr. Whytehead writes:

> "The fact that many goats wear tassels has always interested me, and my curiosity was intensified one day when, on a farm in Dorsetshire, I came across a number of pigs wearing tassels very similar to those of the goat. Some time later, when visiting Armstrong College, I saw some specimens of fossilised fish, of which the openings of the gills were specially drawn attention to. Knowing the supposed connection between tassels and gills, I made enquiries of the officials of the College, and received from their staff the following particulars: Firstly, the fossil evidence relating to gills and almost all organic organs of the prehistoric animals—especially land animals—is extremely unsatisfactory, and practically no development can be traced. Secondly, there is definitely a relationship between the goats' tassel, the growth on the neck of the pig and similar things on other mammals, and the gills of their distant ancestor, the fish. . .
>
> "Fish gills are developed in all Mammalian embryos and the gill slit remains in every case, slowly becoming the ear cavity. In most cases the other gill slits disappear externally, and the neck is perfectly smooth. In cases of embryonic disease, however, the slits, even in human beings, may remain open, or may leave scars which remain throughout life.
>
> "In the case of certain breeds of goats and pigs, the one pair of gills have developed into ears, and the other pair, instead of disappearing, have developed into tassels."

This is the best explanation that I have as yet met with.

Pausanias tells us of a curious use to which a number of goats were, on one occasion, put. The Sicyonians collected an army to invade the country of their neighbours, the Ægeiratæ (the Hyperesians of Homer); but these latter people, not considering themselves a match for the Sicyonians, adopted the following stratagem:—They collected together all the goats in their land, fastened torches on their horns, and at night-time set fire to them. The Sicyonians, imagining that these lights proceeded from the auxiliaries of the Ægeiratæ, led their army home. It was from this circumstance that the ancient name of Hyperetia was changed to Ægeira, so called from *aiges*, the Greek for goats. In the place where the most beautiful goat, the leader of the rest, lay down, the people built a temple to Artemis Agrotera, "the huntress," for they considered that the stratagem which saved them was effectual by the aid of Artemis. During the prevalence of an epidemic or pestilence a goat was sometimes sacrificed to the rising sun. The people of Cleonæ, an ancient town of Argolis, did this, and when they were freed from the plague, they sent a brazen figure of a goat to Apollo (x, 11).

In Æsop's fables the goat is generally represented as being not over clever, as in that of the Fox which had fallen into a well, and the Goat by whose assistance Reynard got out. The Fox remarked to the Goat, whom he left in the well, "If you had half as much brains as you have beard, you would have looked before you leaped." Against this I will give what Pliny, on the authority of an eye-witness, Mutianus, relates as to the intelligence of the goat, and with this little story will conclude this chapter :—

"Two goats coming from opposite directions met on a very narrow bridge, which would not admit of either of them turning round, and in consequence of its great length they could not safely go backwards, there being no sure footing on account of its narrowness, while at the same time an impetuous torrent was rushing rapidly beneath ; accordingly one of the animals lay down flat, and the other walked over it."

CHAPTER III

VARIETIES OF GOATS

THERE is, probably, no animal (the dog excepted) that has a greater variety of range than the goat. It is met with in most parts of the world, and appears as much at home in the cold regions of Norway and Sweden as in the hot countries of Asia and Africa. These varieties differ in the length, colour, and texture of the coat, the shape and size of the ears, the contour of the face, and the configuration of the horns; some having a strong resemblance to the Ægagrus or wild goat, whilst others approximate more in appearance to the sheep. Bénion, a French writer, on the authority of an "Encyclopædia of Natural History," classifies the varieties into four distinct groups: (1) Goats with short prick ears, as the common goat, the hornless goat,[1] and the dwarf goat of Guinea. (2) Goats with long, wide, pendulous ears, as the Syrian, Nubian, or Egyptian. (3) Goats with drooping ears and curly woolly coat, as the Angora; and, lastly, those having semi-pendulous ears and a downy undergrowth of wool, as the Cashmere. Huart du Plessis, another French writer, without grouping them exactly in this manner, simply recognises four distinct breeds, which he describes as the common goat, the Angora, the Cashmere, and the Nubian. Neither of these views appears to me correct. There must be either *two* distinct breeds *only*, or there are considerably more than four. The breeds I refer to are the Cashmere and Angora representing the wool-bearing races, and the other goats whose coats are composed chiefly of hair. This is the great distinction, but, in the case of the Angora at least, not the only one, as I shall point out when

[1] It is generally admitted by writers of the present day that there is no such breed.

describing this animal. At the same time, as regards the nature of the coat, it is a difference which is not so much one of kind as of degree, since all goats possess the two coverings, wool and hair. There is a downy undergrowth, and sometimes of considerable quantity, in our own breeds, as may be easily observed on examination; it serves as a protection against wet and cold, and is present, I believe, to some extent in all quadrupeds, thus corresponding to the down between the feathers of birds.

M. Joseph Crepin, a more recent authority than those quoted above, in his book on "La Chèvre," published in Paris in 1906, devotes a chapter of some 140 pages to "The Breeds of Goats throughout the World," and as he has applied himself for many years to the study of this subject, and has had special facilities for acquiring information upon it, I shall not hesitate to quote at some length from his valuable work. This writer introduces the subject by remarking that "The European breeds are so mixed up that it would take a long time to disengage and reconstitute the types proper to each country. Do we not find," he asks, "side by side in every breed some horned and others hornless, some long- and others short-haired, and, again, some self-coloured and others polychromatic? . . . . The difficulty met with in making such distinctions has led the public to form the most erroneous ideas in appreciating and classifying the caprine race. Do we not constantly hear people talk with an air of assurance of 'white breeds,' 'black breeds,' 'long-haired breeds,' 'horned' and 'hornless' breeds as if the fact of being white or black, of having hair more or less long, or a head garnished or deprived of horns, had ever characterised the origin of a European race." All this is perfectly true as regards the goats of Europe, but, as the author here quoted further observes, "this confusion of races is not so noticeable amongst Asiatic or African varieties, the characteristics of these being much more pronounced." Whilst stating, however, as here shown, that the goats of Europe generally are difficult to divide into breeds, M. Crepin adds that "the disposition of the ears, the shape of the horns—when there are any—and the colour of the eyes [!] are the only means of recognition for an experienced connoisseur." The points of distinction here given are of a singularly meagre character, and if we leave out "the colour of the eyes"—a feature I certainly never before heard mentioned or personally noticed as one which assists in defining a breed or variety—we have nothing left but ears and horns to guide us, and the latter vary, we know, a great deal even in specimens of the same kind.

Now, the goats distinguished as breeds by M. Crepin are chiefly the following: The Alpine, the Schwarzhals, the Pyrenean, the Massif Central, the Corsican, the Murcien, the Grenada and La Mancha of Spain, the Maltese, the Arabian, the Crimean or Caucasus goat, the Bosnian of the Balkans, the Syrian, the Cashmere or Thibet goat, and, finally, the Angora. Illustrations furnished of these different kinds do not in some instances show very marked features of distinction, and these latter are not apparently governed by ears and horns. German writers extend this list considerably, giving names of varieties that have probably only local notoriety. Such are the Saxony goat, the Westphalian, the Hinterwald, the Wiesenthal, the Starkenberg, the Langensalzaer or Thuringian, the Schwarzwald, and the Hartz Mountain goat. Prof. Anderegg, a Swiss authority, credits his own country alone with possessing no fewer than sixteen breeds; but as he treats as one of these the Appenzeller, which is merely a sub-variety,[1] it may be assumed that other of his so-called breeds are but local varieties. All recognised authorities agree that there is, with but slight modification, a common type of goat met with all over Europe, and of late years in America and Australia. This common breed has been crossed in certain countries or districts with imported stock from other countries; or specimens showing some special feature or colour have been selected and bred for a certain period until a fixed type has been established, by which time the animal has received the name of the district or one suggested by its markings. Thus we find the Schwarzhals, meaning "black neck," which has now developed into a breed, and more recently the "Coublanc" or "Couclair" of Crepin, a variety of Alpine which, as its name implies, has a "white or light-coloured" neck. As a remarkable illustration of a nominal "breed" we have the "American Goat." There are no goats indigenous to the United States, and until the latter part of the eighteenth century there were but few of these animals in that

[1] Another Swiss authority says it is a smaller Saanen.

part of the American continent. But interest in this class of stock having been evoked, presumably through the circulation of English literature on the utility of the goat and its increasing popularity in this country, the subject has now been taken up in the States in that go-ahead, enthusiastic manner so characteristic of Americans, and we read in Thompson's "Information Concerning the Milch Goats," published by the Bureau of Animal Industry at Washington, that "the American goat is a name suggested for a breed which it is desirable to develop by selection from the so-called common goats now in this country. It is known that among these goats there are often found some excellent milkers, although their origin is obscure... These are the kind of goats that should be selected as a foundation for the American breed, and if their milk characteristics were further increased by crossing with either the pure-bred Toggenburg or Saanen bucks which are now in this country we should soon see a breed that would produce a satisfactory amount of milk and at the same time have all the hardiness possessed by our common goats." This is quite right, but it will be seen that the animal here mentioned as a breed is at present an ideal one and awaiting development. It is, therefore, rather curious to find in this book an illustration depicting a group of white goats of a certain uniform type with the title: "American Milch Goats." Similarly we find an illustration of a herd of "Australian goats" with drooping ears, though we know that the goat is no more indigenous to Australia than to North America, and goat-keeping has probably been introduced into Australasia only during the last decade.

From the foregoing remarks it will be seen that I am not in favour of grouping the breeds into sections, as adopted by the earlier French writers, as I fail to see how this can be consistently carried out. I shall therefore make no attempt at classification, but content myself with describing the breeds and varieties of each country possessing features of interest and points of distinction.

CHAPTER IV

BRITISH MILCH GOATS

For the same reason as that already given with regard to the goats of Europe generally, there exists a certain diversity of characteristics amongst the goats of the British Isles. These have in times past been crossed with foreign specimens, chiefly from India and Malta, that were landed on our shores from mail steamers and private yachts, having been used to supply milk during the voyage. In later years, for the purpose of obtaining fresh blood and breeding superior milking stock, importations have been made in addition from Switzerland and the Cape, whilst a few Nubians and others of similar type have found their way into this country. All this foreign blood has left its mark in one way or other, and it can generally be detected either by the shape and size of the ears, by the configuration of the head and horns, or by the colour and markings of the coat. Nevertheless two types of the common goat exist, the one in England and the other in Ireland and Wales, which are so distinct that there is no difficulty in identifying them.

The English Goat

I have been at some considerable pains to ascertain the points of the original English breed ; and from the descriptions in old works in which goats are mentioned, and a comparison with the features most prevalent at the present time, I believe the following description to be correct :—Head neat and tapering, with moderate beard ; frontal-bone prominent, horns set far apart, rising slightly at first with an inclination to the rear, and then branching outward ; ears rather small, erect more or less, but without any tendency to break or to hang downwards. Body long and wedge-shaped, with a fairly close coat. In the male, a grand specimen of which is shown in the illustration of "Monster," a prize-winner in his day, the hair is longer, particularly at the neck, chest and thighs, where it is thick and stiff. A fine, soft, woolly undergrowth is often observable between the hair. The colour ranges from black to white, but is more often light or dark fawn, with a darker line along the back, and black on the legs.

For some years past a section of goat keepers, led by Mrs. Walter Paget, have been concerned in a movement to revive the original English Goat as a breed and to adopt measures for its improvement without admitting any Swiss, Anglo-Nubian or other foreign blood. An "English Section" has been introduced into the Herd Book, the condition of entry being: "When the goat itself and its sire and dam, grandsires and granddams are entered in the Probationer's Record" or "When both sire and dam are entered in the English section of the Herd Book." Entry in this Record is governed by Regulation 54: "Goats, before being entered in the Probationer's Record of the English Section, shall be inspected by two or more recognised judges present at the same time, who shall report as to the eligibility or otherwise, of the goats inspected. No kid under the age of six months can be inspected."

Wild goats are still met with in parts of Great Britain. We hear of white Cheviot goats, and mention has been made to me of wild white goats in the Isle of Mull, N.B., which are considered a great nuisance, because they interfere with deer stalking. They are said to be keener scented than deer and more on the alert.

Irish and Welsh Goats

The varieties of Ireland and Wales, and especially the latter, present more the form of the mountain goat, for it may be noticed in many countries that the goat of the hills differs somewhat from that of the plains. In Ireland the opportunities for a mixing of breeds by foreign importations have been much less favourable, and doubtless in the more remote parts the type met with to-day is the same as that which has existed there for centuries.

Fig. 1
TYPE OF THE ENGLISH MILCH GOAT

The Irish goat is quite different from the English ; the hair is long and shaggy, generally a reddish black and white or yellowish grey and white. The head, instead of being short and tapering, is long and ugly, the muzzle being coarse and heavy, with a considerable amount of beard even in the females. The horns are large and pointed, situated close to each other, and extend almost perpendicularly, whilst inclining to the rear. Those of the male goat especially attain sometimes an immense length, a pair in my possession measuring each 30in. Besides the increase in size, they branch out more than those of the female. The size and shape of these appendages render both male and female formidable antagonists when pugnaciously inclined, which they not infrequently are, both to each other and to persons who are strangers to them, so that they are not altogether safe with children. The Irish goat is occasionally a good milker, though its period of lactation is usually limited to about six or seven months, and the yield is comparatively poor in quality. The udder is generally long and narrow, with big teats.

The Original Goat of Wales

Droves of Irish goats used to make their appearance annually throughout most of the market towns in England and Scotland, the herdsman, who was always a thorough Hibernian, calling them "Welsh," though for what reason I have never been able to discover, unless it was on account of their having been landed in Wales on their arrival from Ireland.

The original goat of Wales is described in old works as a large kind and mostly white. At the present day, however, these observations do not apply. In many points it resembles the Irish goat, but is smaller and more symmetrically shaped, the head and horns being lighter and more graceful. The few specimens of this breed to be met with at the present day are not of much value for milk ; the udder and teats are usually small, and, like the Irish goat, it does not remain any length of time in profit.

THE ANGLO-NUBIAN

Whilst mentioning varieties that are common to Great Britain something must certainly be said of the Anglo-Nubian, a name given to describe what was not so much a cross between an English goat and a Nubian (as the term naturally suggests) as a mixture of the common short-haired, prick-eared goat of the country and the Oriental lop-eared breed, be it a Nubian, Egyptian, Abyssinian, Chitral or other similar Indian variety. Such goats came over from India on board P. and O. steamers, and the short, sleek coat, small horns, rich black-and-tan colour, and drooping ears which characterised the progeny of this combination soon made it popular, especially at exhibitions, where most of the prizes have been won with it from the earliest days of goat-showing. Like the Orpington fowl, which we know to have been made up of several varieties and which is now a breed in itself, the Anglo-Nubian has ceased to be regarded as a cross-bred animal, many of those of the present day having pedigrees which date back thirty years and more.

It may be of interest to give here an account of the evolution of the Anglo-Nubian through its successive stages. It originated under the term "English and Abyssinian" with Dr. Crisp, of Walworth, and a Mr. W. Freeman, of Wandsworth, some of whose stock were exhibited at the first and second goat shows held at the Crystal Palace in 1875 and '76 in the name of J. Hahn, by whom they were described as "English and Indian."[1] At the former show a selection was made of some prize winners by the present writer—who bred them for a number of years—for the late Baroness Burdett-Coutts, who was then starting her subsequently famous herd. From 1878 to 1880 Dr. Crisp himself exhibited and won at the Dairy Show first prize, as well as an extra milking prize of £5, with a goat which was described (as I find on referring to my report in *The Bazaar* of that time) as "a nearly perfect specimen of

[1] This is according to my own knowledge, but I have since been informed by Lord Arthur Cecil that, in 1859, as a child he used to drive about the park at Hatfield a lop-eared goat which was known as "the Syrian," a breed with very similar features to the Nubian; so it is probable that the crossing of the Eastern with the common breed of this country originated some years before I had cognisance of it.

the Nubian breed, tall in the extreme, with long legs, Roman nose, very close coat, and long pendulous ears."

Dr. Crisp dropped out of goat-breeding after this and died a year or so later, but the Anglo-Nubian was for many years after bred successfully by Mr. W. Freeman and his man Mr. Dormer, both of whom won many prizes with them. In 1880 the late Mr. B. Ravenscroft took up this variety, which he bred consistently for years, owning at one time the largest herd in England. Some few years later this breed came into the hands of Mr. C. L. Jackson, of Bolton, who figured largely as a prize-winner at all the Midland shows of those days, as well as subsequently at those in the South, until he disposed of his complete herd to the late Sir Humphrey de Trafford in 1895.

Mr. Sam Woodiwiss was the next to come on the scene, and at the nick of time to revive the Eastern characteristics, which had begun to wane for want of fresh foreign blood, there being by this time more "Anglo" than "Nubian" about the breed. On the first opportunity this enterprising breeder, of bulldog fame, obtained through Mr. Jamrach, the well-known importer, that so-called "Nubian" (it was really a Jumna Pari Goat. See "Indian Goats"), Sedgemere Chancellor. There was at once a rush by goat-breeders for this new blood, and its effect was quickly apparent. Mated to the Baroness Burdett-Coutts's Gertie, a she-goat of wonderful size and shape besides possessing great milking properties, they produced Grey Rock, the most popular and most sought-after winner in the Prize Record. With his progeny the Anglo-Nubian fanciers got back the arched face (commonly referred to as "Roman nose"), delicate nostrils, and long drooping ear displayed in the early crosses. Grey Rock, mated in 1897 to the present writer's London Pride, produced Marmora; this was sold to Mr. Ravenscroft, who with it won a number of prizes, and used it largely in the production of his many prize-winners. Later Marmora went to Killerton, near Exeter, as the property of Lady Acland, who attributed to his blood the improvement of her late herd. Among the principal breeders and exhibitors of the Anglo-Nubian in later years were Sir H. F. de Trafford, Bart., the Hon. Mrs. Pomeroy, Mr. and Mrs. Reginald Pease, Mrs. Picard, Mr. Horn, Miss Pelly, and more recently Dr. B. D. Z. Wright. This breed has become much less popular of late.

Other importations will be mentioned when dealing with Nubians. A Chitral male called Bricket Cross was introduced by Mr. B. Ravenscroft about the year 1904, which supplied again fresh blood for this once popular variety. It is much to be regretted that the Anglo-Nubian as a breed is depreciating for want of new blood, but steps are being considered with a view to the importation of fresh specimens from abroad.

The points of a typical Anglo-Nubian are as follows: Coat short throughout, with no fringe of long hair on the back or long tufts on the flanks; colour, preferably black and tan, or reddish-brown with or without black or black-and-white markings, but free from white streaks on the sides of the face, which would indicate Swiss blood. The horns, if any, should be small and curve downwards and outwards rather than directly upwards. The ears must be long, wide, and pendulous or semi-pendulous, but not broken or twisted. The facial line should be plainly arched, the head neat, with a slight taper towards the muzzle, which is small, and in the female without beard. The eye should be large and full, and the forehead wide.

The Anglo-Nubian, in addition to its attractive appearance, is a good milker of rich milk, containing more butter-fat than that of Swiss goats, although the yield is not as a rule so abundant. It grows rapidly and attains a large size. For draught purposes this breed is unequalled, especially the bucks, for these when emasculated and well groomed have close, glossy coats like those of horses, and look extremely well in harness. The male kids are generally plump and meaty, and when killed within a month old make a capital dish for the dinner-table. Specimens of the Anglo-Nubian have been imported into the West Indies, and thence to the United States of America.

Other Recognised British Varieties.

As a sequel to the importation of goats into England in 1922, the British Goat Society recognised three other varieties resulting from crosses with various Swiss breeds. Viz: The British Alpine, The British Saanen and the British Toggenburg.

The British Alpine. The British Alpine is essentially a "Swiss" in appearance, and might be described as a black

Toggenburg. The same objections in the way of foreign characteristics which would tell against a pure Toggenburg would be equally applicable to this variety.

Points: A long lean head, with or without beard, horned or hornless. Facial line straight and muzzle somewhat coarse, as in the Toggenburg and the Saanen. Ears erect or pointing forward; neck long and slender, with or without the tassel-like appendages common to the Swiss. The body colour must be black, with or without white on the belly, together with two white or light fawn stripes along the face, white or light fawn on legs from knees and hocks downwards, and white or light fawn at the rump and on or about the tail—a little white or light fawn about the cheeks or under the neck is permissible. White or light fawn patches on the body or forehead are not permissible, whilst goats with dark fawn, brown, or tan markings are not acceptable. White markings are preferred. The coat should be short throughout, but there may be a ridge of longer hair along the back, and a thicker growth at the hind quarters.

THE BRITISH SAANEN.—Points: A long head, facial line straight with coarse muzzle, with or without beard, horned or hornless. Ears erect or pointing slightly forward, but never broken or pendant; neck long and slender, with or without tassel-like appendages. Colour, white, pale cream or very pale biscuit is accepted. Black skin spots are allowed. (Goats with black spots on the hair should not be accepted.) White goats are preferred. Coat should be short, with or without a fringe down the back and down the hindquarters.

THE BRITISH TOGGENBURG. General Description: The British Toggenburg is essentially "Swiss" in appearance. The same objections in the way of foreign characteristics which would tell against a pure Toggenburg—with the exception of colour—would be equally applicable to this variety.

Points: A long lean head, with or without beard, and with or without horns. Facial line straight and muzzle somewhat coarse. Ears erect or pointing forward, but never broken or pendant. Neck long and slender, with or without tassel-like appendages. Colour may be anything from light drab to dark chocolate, with or without white on the belly, together with two white or light fawn stripes along the face, white or light fawn on legs from knees and hocks downwards, and white or light fawn at the rump, and on or about the tail—a little white about the cheeks, or under the neck is permissible. White patches on the body or forehead are not permissible, whilst goats with dark fawn, brown, or tan markings are not acceptable. White markings are preferred.

CHAPTER V

CONTINENTAL MILCH GOATS

THE GOATS OF FRANCE

AN inquiry was instituted in 1883 by the Société Nationale d' Acclimatation of Paris on the subject of goat-keeping in France, and various questions were drawn up and sent to correspondents in the different departments, having reference to the varieties kept in each, the methods of treatment, utilisation of products, and other matters of interest. The result was embodied in a report drawn up by Monsieur J. Gautier and published in the Journal of the Society. From the answers to questions put, it would appear that there is even greater variety among the goats of France than of Great Britain; but at the same time there exists no particular French breed (though M. Gautier himself is of a contrary opinion). The characteristics differ chiefly in those departments bordering on the Alps and Pyrenees, the goats of these mountains being of quite a distinct type from the rest, as also from each other. If we except these goats (which belong perhaps as much to the countries bordering the Alps on the other side as to France) and one other which I am about to describe, we do not find even in M. Crepin's book any absolutely French *breeds*, though he mentions many *varieties* resulting from crossings between the Alpine and Pyrenean and the common goat. Amongst these, though raised to the dignity of a "breed," is the goat of "Le Massif Central," the type of which this author admits to be "badly fixed," and regards as "resulting from a cross between a degenerate Pyrenean and the common goat found in every latitude." This animal, in the illustration in the book referred to, has indeed all the appearance of a common hornless Irish

C

goat, a similitude which is further borne out by the description. We learn that this goat, which is met with chiefly in the neighbourhood of Poitou, is meagre-looking, with short prick ears and a straight forehead, but finer at the muzzle than the Alpine or Pyrenean. The face is streaked on each side with a white line from under the eyes to the nostrils. "The colour is a tawny brown, or black, the hair on the body long, especially on the back and hindquarters, but short at the neck and head, though this varies with different specimens, some having hair of medium length and others quite short"—corresponding in fact, to the common goat.

THE PYRENEAN.—We have it on the authority also of the breeder from whose book I have just quoted that in the Pyrenean chain there are many varieties of goats, but from what he tells us these varieties appear to be a heterogeneous lot made up of crossings between the original mountain breed and various other breeds from neighbouring countries, whence they have either wandered or been imported, including those of Spain, Malta, Algeria, and even Thibet. The Pyrenean goat proper, however, is the Bearnese, a breed we have seen specimens of in England, some of which in 1880 were exhibited and crossed with our Anglo-Nubians, but of these I am unable to give a very satisfactory account. If size and bone were the principal objects in view in goat-breeding I should recommend using the Pyrenean as much as possible, for those imported specimens were undoubtedly the largest she-goats ever seen in this country. Great size, however, and a corresponding capacity for disposing of a vast amount of food without giving an equivalent return in the way of milk, are not features to be sought after, especially when further handicapped by the introduction of long hair and tremendous horns. There are goats, I believe, in existence still descended from those imported Pyreneans which yet bear traces of the coarse horns and large bones of their ancestors, although most of the long hair has been bred out of them.

It is perhaps hardly just, however, to condemn a whole breed on the limited experience we have had in this country with a few imported specimens, and as M. Crepin speaks in very different terms of this goat I will proceed to give his description and opinion of it: "There exists in the environs of Lourdes a superb goat generally called 'Chèvre du Bearn,' which appears admirably adapted to make a milker of the first order. Its hair is long, brilliant, and silky, and generally black with white under the body and on the feet and legs up to the knees. It is also, however, found entirely white. It has a powerful head, with a straight forehead, ears heavy, and a full eye of a light brown colour. A network of disordered tufts of frizzled hair is displayed over the forehead, giving a remarkable expression of vivacity and elegance. Limbs delicate and well set on. The gait is brisk and alert, and the disposition gentle. It is a pretty creature of great height, giving in full milk as much as 4 litres a day and preserving its lactation for a considerable period. The milk has an exquisite taste without the least vestige of caprine odour, and contains as much as 50 grammes to the litre of butter-fat, equal to that of the Murcien." This seems to me a highly flattering description, and one which is not altogether borne out by the appearance of the animal in the illustration supplied in the book referred to, where it has a decidedly dejected appearance and gives no evidence of being a remarkable milker. Large herds of these goats are driven through France, and even as far as Belgium, in the same way that Irish droves are distributed about England during the summer months. In years past, before the restrictions against the importations of goats from abroad, some of these Bearnese goatherds found their way into this country, and even into London, where the sustenance of their animals seemed to be chiefly derived from eating the advertisements off the hoardings. It was from these droves that the specimens I have alluded to were obtained.

THE SUNDGAU.—This goat, a variety of the Alpine, is said to be of Alsatian origin. It has short hair of a brilliant black with white markings about the tail and under the body and on the legs. White is also more or less prominently displayed on the head, sometimes in streaks of various depths from the top of the forehead down to the muzzle, sometimes in a mass all over the face. The illustration of Mr. S. Woodiwiss's Sedgemere Faith so perfectly agrees with both the description and illustration in M. Crepin's book that I use it here to illustrate this breed in preference to the Alpine, which this goat has been generally supposed to be. There is a further agreement between Sedgemere Faith and the breed I am here describing, inasmuch as the former, which won many milking prizes, gave under careful observation the largest yield then recorded in this country whilst Crepin credits his specimen of the Sundgau goat named Frenélé as being the heaviest milker in his experience, having given the incredible quantity of 8 litres (over 6 quarts!) in a day. The breeders of the Sundgau, we are told, obtain high prices for their animals, the females fetching from £4 to £6 each and the males very much more. Many are purchased by the Government, which distributes them in different districts for the improvement of local strains.

THE TARENTAISE.—This is another variety claiming Alpine origin, for it cannot be called a distinct breed by any means, as its peculiarity is restricted to its colour, which, like the Schwarzhals amongst Swiss breeds, consists of two kinds, In this case, however, it is the head and neck that are light. being saffron-red, and the body black, with a black bar down each side of the face. The ears are long and decidedly erect, the coat is close and shiny, the body long, and the head, like those of other Alpine varieties, neat and compact. Several specimens were on view at the Paris Goat Show in 1903, and they were regarded by English judges as almost, if not quite, the most attractive collection present. These goats are to be found in the Tarentaise and other districts of Savoy bordering on the French Alps, but they are not very generally met with.

THE GOATS OF SWITZERLAND

Switzerland, of all European countries the most adapted to the goat, seems to have devoted the greatest care and attention to the breeding of this animal for milk production, and this has no doubt been going on for a vast number of years, as breeds like the Toggenburg are very prepotent and fix their characteristics on their progeny in a very marked manner, even when crossed with other kinds of a quite different type. The French authority mentioned is, however, very loth to give the Swiss credit for their good work in this direction, for he claims as French such goats as the Toggenburg, Saanen, and Alpine, regarding the first two as mere varieties of the last.

Thus we read in his book under the heading "Race Alpine": "We have already said that all the caprine varieties that we possess in the French Alps exist equally in Switzerland, where they know how to set a value on them under the denomination of breeds. Thus it is that all white or cream-coloured goats without horns and with short hair are characterised as Saanens." Now, such a remark as that first quoted may apply to Alpines, but it is difficult to see how it can be applicable to the Saanen or Toggenburg, for though these animals are kept in France, as they are in Belgium, Germany, and other countries, their home is, and has always been, in those Swiss cantons whence many of them derive their names. We further read: "Our neighbours breed specially with a view to colour, and apply themselves to abolish horns. They think to be able thus to transform varieties into actual breeds; but we affirm, basing our opinion on a long and attentive experience, that there exists still no difference in the milking value of a selected Alpine from Switzerland and one equally selected from France." In the previous paragraph to that from which these words are extracted we are told that "in Switzerland the goat is better treated and better cared for than in France." Now we all know, and M. Crepin himself admits, that it is this extra care ("*bons soins*") that a goat receives which makes it a better milker than others not so treated, and especially does it tend to a long continued lactation, both these features being particularly observable in most of the goats of Switzerland, and especially in those that are well looked after. If, therefore, this judicious system has been pursued for a great number of years there is every reason why Swiss breeds should be better milkers than those of France.

No better testimony to the value of Swiss breeds of goats could be afforded than the exportation which goes on annually from the various cantons to every part of the civilised world where a demand for milch goats has been created. This has been the case especially during the last fifteen or twenty years, and the consequence is that prices have considerably risen since the first consignment of Toggenburgs was landed in this country.

THE ALPINE.—According to the author of "La Chèvre," the pure Alpine should possess the following points: "Neat head, wide at the forehead, with a slight prominence over the eyebrows. Eyes, well formed, light brown in colour, and with a gentle expression. Ears erect and of medium length, but horn-shaped and sometimes rather heavy. The muzzle is well pronounced, as it should be in a large feeder. The body is long, with a well-developed stomach. The coat is short, allowance being made for a fringe of hair over the spinal column and on the thighs. The colour is variable, with different shades even on the same animal, and is strongly marked by streaks or washes in all the tones of black, grey, chestnut, fawn, and even a brown-mauve. The Alpine develops rapidly and soon attains a good size, measuring from 31in. to 37in. in height and weighing from 154lb. to 198lb. The length of the body from the head to the root of the tail is, in the she-goat, from 45in. to 47in. In the male, which is often nearly 40in. at the withers, the hair is much longer and closer, covering the greater part of the forequarters." Whilst this writer regards the Alpine as the pure breed, and the Toggenburg and Saanen as varieties of it, German and Swiss authorities class these latter only as pure, and the Alpine as a nondescript common goat which seems much more correct.

Mr. G. F. Thompson in his American treatise on goats (previously referred to), quoting from Mr. Peer, who, he says, made no little effort to learn something about it, writes: "The Alpine goat may be termed the native goat of Switzerland, corresponding in name to our 'Common' goat and the German 'Farm goat,' but not corresponding in important characteristics. These goats are found throughout the whole Alpine chain. There is but little uniformity amongst them, and therefore they can hardly be classed as a distinctive breed. They resemble grade animals of nearly every description. . . . In some districts in the Swiss Alps, and also in the French Alps some attempt has been made to breed to a certain form and colour, but, as a rule, they are better classed as 'all sorts.' "

These remarks entirely agree with the observations of English travellers in those regions.

THE CHAMOISEE, OR GUGGISBERGER.—If a distinct Alpine breed exists, it is probably the "Chamoisée," a name given to it by the French on account of its having some resemblance chiefly in colour, to the Chamois. M. Julmy, professor at the

School of Practical Agriculture of Écone (Valais), in his monograph on "The Breeds of Goats of Switzerland" (from which I obtain most of the information here given), includes this amongst the four principal varieties of the Swiss Alps, the others being the Schwarzhals, the Gessenay or Saanen, and the Toggenburg. According to this author the Chamoisée is an old-established mountain breed, and one to which many comparatively modern varieties, including the Toggenburg, owe their origin. Amongst these no doubt are the Schwarzwald of Hoffman, the Gruyère, the Sundgau, and the Tarentaise of Crepin. This is the more probable inasmuch as the breed in question is one of the most widely distributed throughout the Swiss cantons.[1] It has, in consequence, various names according to the locality. In the canton just mentioned it is called the "St. Gallen Oberlander" or "Stiefelgeissen," in the Canton des Grisons "Einheimische Rasse, braunchwarz," and in the Unterwald "Obwaldnerschlag braun" and "Gemsfarbige Ziege." All of these names have reference to its brown colour, which pervades the coat generally, with markings of a darker shade along the back extending to the tail, the face, ears, chest, and shoulders. The head is short and lightly set on, the forehead being somewhat depressed. The horns (in the female) are slight, and take a more or less vertical direction, separating as they rise with a rather sharp curve downwards at the extremities. The hair is short and thick on the body, but longer on the back and thighs, a feature noticeable in most breeds that are much exposed to the weather.

This goat is but little sought after for exportation, but Crepin has introduced a considerable number into France, and it is very popular in Belgium, where it is being encouraged in the establishment of an improved Belgian breed.

Mr. G. F. Thompson, in his American work, "Information Concerning the Milch Goats," already alluded to, makes mention of the Guggisberger goat, which he says, "is sometimes called the 'Schwarzenburg-Guggisberger,' and was originally from the Simmen Valley." We are told that "it is brown, spotted like the chamois, commonly has horns, and is con-

[1] It is doubtless this goat that Mr. Bryan Hook refers to in his "Milch Goats and their Management" as "the common reddish-brown mountain goat of the Upper Wallis."

sidered very excellent for milk" ; also that Hilpert (presumably a German writer, as Thompson gets his information largely from German authorities) describes it as being built like the Saanen—with which it has been extensively crossed of late years—"but of fawn colour or brownish white and of astonishing size," and says that "it is excelled by no other breed of goat in milk production when under good care and feed." It is finally said to be "a near relative of the 'Oberhaslian goat.' " A picture of the Guggisberger is given in Thompson's book, which shows a very long-bodied, even-coated "doe," hornless, and with a large spherical udder. There can be no doubt, however, that the Guggisberger is only another name for the "Chamoisée," the animal depicted in the book just mentioned being of precisely the same type, but without horns.

THE TOGGENBURG.—This breed is said to be the result of a cross between the white Appenzell and the Chamoisée, both of which it is gradually supplanting at the present day. Although most goat-keepers on the Continent place the Saanen in the foremost rank among Swiss breeds, English fanciers—their experience with the Saanen being comparatively recent—give preference to the Toggenburg. This breed is gaining also in favour in the United States, into which country both kinds have of late years been introduced. This goat, as we know it and breed it in England, is a short-haired animal, though in the Canton St. Gallen, where the Toggenburg has its home, a very large proportion are met with having long and almost shaggy coats, especially in winter. It is also supposed very generally to be a hornless kind, and, indeed, in Switzerland polled specimens are the rule rather than the exception, for the simple reason that the kids that are born with horns are usually killed and eaten. Professor Julmy says:

"The true Toggenburg should never have horns ; when these occur they are cases of atavism, which are becoming more and more rare." This is plainly shown in the illustration of a herd photographed by a visitor to the Toggenburg Valley. These are typical specimens and all hornless. In England this breed is too rare and too valuable for its patrons to destroy thus any pure stock, and consequently with us horned Toggenburgs are far from uncommon. This goat enjoys a unique distinction in the matter of colour, being uniformly of a light or occasionally rather dark drab, best described as mouse-colour, with white or greyish-white markings. The markings are thus distributed: There is always a streak on each side of the face from right up under the ears down to the muzzle. White is also displayed on each side of the tail, under the body, and sometimes on the ribs, but always on the legs between the thighs, and from the knees downwards. The head is rather long, the facial line being straight or slightly concave; the eye is full, with a gentle, kind expression; the ears are of medium size and more or less erect, though held sometimes horizontally. These also have white inside and at their base. The coat in the female should be fairly thick, but the hair should be soft and, as already stated, short, or at most of medium length. The neck is long and slender, and there should be always the two tassel-like appendages observable now and again in other breeds. There is generally a beard in the female, but in the male it is long and heavy. With him also, as with most male goats, the hair of the body is longer and thicker, especially at the neck, chest, and also on the thighs. On many of the females this extra length of hair is noticeable on the hind-quarters, and frequently it forms a fringe along the back. The Toggenburg is nearly always a lean animal, small in bone, of medium size, with a slender neck, and frequently protruding hip-bones. Although this and its prominent ribs are no fault, these latter, as they should be, are well sprung; indeed, flat-sidedness is a very uncommon defect with this breed.

The average height of the Toggenburg she-goat is from 28in. to 29in. at the withers; but English-bred specimens are generally larger than those that have been imported, the size having greatly increased since they were first introduced. Thus Polidore (a well-known male owned by the late Mr. Paul Thomas, to whom the credit is due of having first imported this breed into England in 1884) scaled 117lb., a weight greatly exceeded in later years by Mrs. Straker's Kaiser, which in 1906 weighed 137½lb., and more recently by Mrs. Handley Spicer's Copthorne Nimrod, the largest and heaviest male Toggenburg that has been yet produced in this country. This very fine specimen at two years old weighed 181lb., his height being

35in. (This goat died, unfortunately, on his return from Tunbridge Wells Show in July, 1909.) Similarly Galatea, a first prize she-goat, weighed in 1893, at five years old, 97½lb.; whilst Mr. Sam Woodiwiss's Sedgemere Belmont in 1906 weighed 115lb., and his Sedgemere Celia, probably the largest doe of this kind ever exhibited, at about the same age weighed 133lb.

The Toggenburg may be said to have been the most popular breed of goat in England for years, as indeed it has been in other countries. This is due to two important qualities—its great milking capacity and its docile, kindly nature. Anyone visiting a herd of goats of various sorts amongst which there are Toggenburgs will generally find that whilst the others move off at the approach of strangers, the Swiss will come up to be patted, and will feed readily from the hand. In regard to its milking qualities a great deal could be written. Amongst those breeders who have done much in earlier days to make this goat popular and who have devoted personal attention to its breeding and milking are the late Mr. Bryan Hook, of Churt, near Farnham, and Mr. Gates of Guildford. Both these breeders have kept records of the yields of their animals, particulars of which may be read with interest. A goat in the herd of the former, although not quite pure-bred, named Uma, a winner of several milking prizes, gave about a ton of milk in two years. Having kidded on 9th May, she gave on the 12th 5lb. 2oz. a day. On 19th June she was giving 6½lb., and so on till 23rd Jan. of the next year, when her yield had dropped to 2lb. daily. In Mr. Gates's herd one goat gave a daily average of 6½lb., another 4lb. 6oz., a third 4lb. 3½oz., and a fourth 4lb. The largest quantity given in one day by one specimen, Melissa, was 8lb. 11½oz. (10lb., it should be observed equals one gallon). Five goats gave during their milking period in 1898 a total of 7,140lb. As a set-off against this productiveness for quantity it must be observed that the quality is poor, the milk of this breed, and indeed of most Swiss varieties, being deficient in butter-fat. The chief reason for the Toggenburg being so abundant a milker consists in the fact that in Switzerland special attention to this feature has been given by breeders from time immemorial, inferior milkers being killed and eaten, and only the good ones kept for breeding.

This process of selection was still further carried out in all the specimens that have been imported into England, for the best prices were paid and only the *crème de la crème* were brought over. Hence it is comparatively rare to find a bad milker amongst the pure pedigree stock in England. In the Herd Book of the British Goat Society a separate section is devoted to this breed, one of the conditions of entry being that the parents are imported or directly descended from imported stock.

More recently some Toggenburgs in the herd of Mr. Sam Woodiwiss, as also in that of Mrs. Handley Spicer—both of whom made a speciality of Swiss breeds—have shown some wonderful performances at the milking bench. Mr. Woodiwiss's Martha gave at Tunbridge Wells in 1906 5½ pints in two milkings; Mrs. Spicer's Trima at the same show in 1908 rather over 9 pints in three milkings. At the dairy Show, 1907, Sedgemere Capella, exhibited by the same lady, who won with her the first milking prize and Baroness Burdett-Coutts's cup, gave in two days over 17lb. of milk, or an average per day of over 7 pints, the goat having been milking 183 days. These yields were given at shows, where goats have not the same chances of milking well as at home.

THE APPENZELL.—This is practically a white Toggenburg, the type being exactly the same, and only the colour different. Prof. Julmy regards it as simply a small variety of the Saanen, yet it is sometimes treated as being a distinct breed. Crepin, however, makes no reference to it in his book, so it is probable that he includes it amongst those referred to in the sentence quoted on a previous page.

When Mr. Bryan Hook visited Switzerland in 1893 for the purpose of bringing home some Toggenburg goats, he was tempted to add to that importation some female Appenzells, which he purchased in various villages, viz., Wildhaus, Unterwasser and Neslau, to secure different strains. He says in his book on "Milch Goats and their Management," after referring to the Toggenburg: "Side by side with this breed and occupying the same district I found another, differing only in colour, and known as the Appenzeller. It is pure white and invariably hornless, and I was informed that large herds are pastured on the snowy slopes of the Säntis mountain, whence the milk is carried on the herdsman's backs to the town and neighbourhood of Appenzell to supply the goat-milk-cure establishments, which are a feature of the little canton to which the town gives its name." The same writer tells us that "as a rule the peasants are very careful to keep the Appenzell and Toggenburg (known locally as the 'white' and the 'red') breeds perfectly distinct, but now and then one sees animals which by their colouring show a cross of the two breeds. All such animals, however, are rigidly excluded from the Government subsidy."

THE SAANEN, OR GESSENAY.—This is considerably larger and coarser than the native breed of the Toggenburg, and, with perhaps the exception of the Schwarzhals, than all Swiss breeds. It takes its name from the Saanen Valley, in the south of the Canton Berne, but is also met with in considerable numbers further north in the upper Simmenthal Valley, where it is known as the Gessenay goat. The Saanen is always white, or at most cream-coloured, and is, like the Toggenburg, supposed to be hornless, as indeed most of them are; but since horns show themselves very frequently even in the progeny of polled parents this cannot be said to be an absolutely hornless breed any more than the one last mentioned. The hair is short, or at least comparatively so, for it is not smooth, and there is generally a fringe of long hair along the back and down the hindquarters. The muzzle is flesh-coloured, and the eyes have a yellow tint with white eyelashes. In the males there is also a tuft of hair hanging over the forehead between the horns. Both males and females have beards, that of the male, as in the case of the Toggenburg, being very long. Other noticeable features of the Saanen she-goat are its long body, slender neck, and long milking-looking head. I have read somewhere that the head of the male is decidedly masculine; but in all the specimens I have seen it was, on the contrary, decidedly feminine, which is generally considered a defect.

The Saanen is highly estimated as a milker in its native valley, and this character has gained adherents for it in several parts of Europe, especially in Germany, France, and Belgium, where it is largely imported. "It is said that in 1893 several thousand head were taken out of the Saanen Valley. Specimens were brought to this country amongst the herd collected together from the Paris Goat Show in 1903, and great things were expected of them, but their reputation as milkers was not maintained, and the breed has with us practically died out." Peer says of this breed: "As compared with the Toggenburg family, my observations lead me to say that there are probably more large-milkers among the Toggenburgers than among the Saanens, but that the best of the Saanen goats are superior to the best Toggenburgers; in fact, that the best of them are probably, the best in the world, giving from five to six quarts a day. One German writer says that some Saanen goats for a time give from 7 to 10½ pints of milk a day, and that this happens not occasionally, but as a rule in the Simmenthal district." We possess, however, goats in England at this date that exceed even these yields.

Professor Anderegg, a Swiss authority, writing of this breed, says: "She is the commonest and one of the largest in the Canton Berne, and thrives as well in summer on the mountains as she does in winter in the stable. By reason of her usefulness she is preferable to any other of the Swiss goats."

THE SCHWARZHALS.—This is a long-haired breed and one that is always horned. Its colour, whence its name "black neck" is derived, is peculiar, inasmuch as it is black on the head, neck, and forequarters, and pure white on the rest of the body and hindquarters, the division of the two colours being sharply defined. The Schwarzhals, known also locally as the Glacier "Goat" and "Saddle goat," comes from the Canton Valais, and is especially numerous in the Rhone Valley as far as Sierre. One writer gives its height as from 28in. to 30in., but says its full size is not attained until it is from four to five years old. All who have seen this breed in Switzerland agree that it is singularly picturesque and handsome, the male especially being a noble-looking animal. He has large spreading horns, a broad forehead, but short head, with, like all Swiss goats, a somewhat coarse muzzle. His beard is of immense length and very thick. The hair is abundant also over his neck and chest, and there is often a tuft on the forehead, with long hair like whiskers on the cheeks. In the female the horns rise up close together with a gradual curve backwards. The ears are sometimes carried upright and sometimes horizontally. This breed is very hardy, as it need be, for according to Mr. Bryan Hook these goats are not so carefully treated and housed as are those of the Toggenburg, and after being milked are driven off to spend the night on the mountain side. Very few specimens of the Schwarzhals have been brought to England, and, in fact, this is not a kind of goat that would find favour with us. The long horns and long hair, and the fact that it is but an indifferent milker, are not points that would commend it to the practical English goat-keeper. Yet it is said a large exportation takes place every year from Switzerland, herds being sent to France, Holland, Italy, Germany and Austria, where this goat is in great repute. The prices of she-goats vary between £3 and £4.

The Goats of Germany

There are a number of varieties of the genus *Capra* mentioned by German writers as peculiar to certain localities, these being named after the districts where such kinds are popular, but in the illustrations given of these animals there seems to be practically nothing in their appearance to distinguish them from the common goat. Thus we read of the Starkenburg as a noble animal in which a large infusion of Saanen blood prevails—a very probable circumstance considering that this Swiss breed is imported in great numbers into Germany.

The Hartz Mountain goat is more likely to be a native race, as most breeds have their origin in wild ancestors whose home was in some range of mountains. The description of this breed is from a German source: "The colour varies from whitish-grey to reddish-brown, with a dark streak along the back; sometimes it is black or brown, or a mixture of these colours. The hide is quite thick—a condition due to the severe climate. The hair of medium length, seldom short and smooth; the head short and broad; eyes grey, ears long and narrow, yet many have lop or so-called mouse ears"—a contradictory statement, it seems to me, as a mouse-eared goat is certainly not lop-eared, and if it were it would indicate an admixture of Eastern blood. The loins and chest are said to be broad, body deep, thighs rather full, legs generally straight, but in exceptional cases cow-hocked. Most of these points are no indication of breed, being features in different specimens of almost every kind of goat. All that Crepin tells us of the "German goat" is that its colour is the same as that of the

Alpine, namely polychromatic; that its coat, although short, has the same fringe along the back and on the thighs as is noticed on many Toggenburgs, and, he might have added, on most common goats. This again shows no points by which a breed can be identified. Similar remarks may be applied to the "Westphalian goat" referred to by Dettweiler, the "Hinterwald" described by Huber, the "Saxony" goat, and the "Wiesenthal" mentioned by Petersen. We may therefore justly conclude that there is no distinct breed of goat in Germany, where Swiss breeds and crosses between these and the common goat are mostly kept.

An article on "The German goat breeding industry," contributed to the 1926 Year Book, by A. Topps, of the "German Goat-breeders' Club," stated that a census taken after the war showed that in New Prussia, there were 2,698,582 goats, and in the German Reich, 4,321,298. This shows that goats in that part of Europe are kept on a very much larger scale than in Great Britain. Some German-Swiss breeds have been imported in considerable numbers from the Saanen and Simmen valleys, as also doe-red Guggisbergers and Toggenburgs, but the Saanen is found to be by far the most popular.

The Maltese Goat

The goat in Malta is an animal of considerable importance, for not only is it the chief source of milk, but it acts the part of the milkman in towns and cities, carrying its own commodity from house to house, even up stairs on different floors, and being pure and unadulterated. The taste of goats' milk on that island, however, differs from that obtained from most of the goats in England, as there it has a decidedly strong and what we should call "goaty" flavour, due, no doubt, to the stale milk which is constantly left in the little cans used for milking the goats decomposing with the heat, and also possibly to odd scraps that the animal picks up in the streets, for the goat in Malta is a dirty feeder.

The Maltese goat is a comparatively small breed, standing about 26in. or 27in. high at the shoulder. It carries a coat of long, straight hair on the body, but on the head and neck the hair is often short and smooth. As with most breeds, the colour varies considerably from white or grey to black, whilst some are spotted. Those I have kept or seen have been generally either greyish-white or reddish-brown. This breed is regarded as practically hornless, but it is not absolutely so, as a few of those in Malta are horned. The ears vary somewhat in regard to their position. Some are carried horizontally, whilst others are what we should call semi-pendulous, and a few completely so, these being extra long and with a slight outward curl at the extremities, showing more or less of the Syrian, from which this breed is supposed by some to originate, in part at least. The facial outline is straight, the head long and narrow and free from beard, this last being a noticeable and special feature of the breed, and imparting to the female a more feminine appearance than have some goats. When horns are present, which is undoubtedly a defect, they are small and curl back over the head. The udder of the Maltese is a long narrow bag, with large, unshapely teats like little udders themselves, its colour like that of the skin generally, being of a peculiar orange-red.

These goats are nearly always thin, however well fed, and often very flat-sided. Like all breeds which have been specially bred for the sake of the produce for a great number of years, the Maltese is an exceptional milker, and being so constantly accompanied by the goatherd, who tends it with care and devotion, it is a very docile and intelligent creature. Crepin says that the goatherds of Malta are in the habit of encouraging the secretion of milk in their goats by a prolonged massage of the udder with well-greased hands during the latter months of gestation. This operation is repeated as often as possible, and the goat not only becomes accustomed to it, but seems quite to enjoy it. The effect of this gentle friction thus constantly applied is to encourage a flow of blood to the mammary glands, which descend and develop towards the lower part of the udder, thus accounting for its singular shape. Another peculiarity we learn from the same source is that the Maltese never completely milk out their goats. Whilst we in England, like the goat-keepers in Switzerland and other parts, regard it as a matter of the highest importance never to leave any milk in the udder when milking, the goatherd in Malta takes a contrary view, explaining that by leaving, say, a wine-glassful behind it keeps up the temperature of the bag and thus encourages the formation of a fresh supply. These people also have an idea that their system of milking small quantities at a time and often, as they do in selling the milk, creates a constant demand on the secreting glands, which Nature responds to accordingly. However much truth there may be in these theories, the fact remains that the Maltese obtain enormous quantities of milk from their goats.

It should here be remarked, as somewhat bearing out this view, that however good as milkers these animals may be on their own island, their yield proves to be comparatively small when they are transported to other parts. Thus, although largely met with in Tunis and Algeria as well as in some parts of Italy and also in Spain, the milk supply from the Maltese goat in the hands of the Arabs and others is not what it is in Malta, and this has been equally the case in England.

Milking in Malta is always performed from behind, and not at the side of, the goat, the long thin-skinned udder, which sometimes nearly touches the ground, naturally lending itself to this method, which could not be conveniently adopted with a spherically-shaped bag.

The Maltese goat years ago, before the law precluded the importation of such animals into this country, was not infrequently seen in England, as a great many were shipped on to yachts and steamers for supplying milk on board, and so brought into this country. The specimens I myself possessed in the early 'eighties were obtained through this means, having been sent to me by the late Lady Brassey, who had always three or four on board her yacht *The Sunbeam.* These goats have also been imported largely into India and a few into America.

Further information on this breed of goat will be found in the chapter on "Goats and the Malta Fever."

Spanish Goats

Spain, like other countries bordering on an extensive range of mountains, possesses a great number of goats, most of which are of the common type. Crepin, however, credits this country with three special breeds, which he says are of Mauritanian origin, viz., the Murcien, the Grenada, and La Mancha, the last differing from the other two in its very small ears. The writer referred to gives the characteristics of the Murcien as follows: "Hair very short, silky, and glossy, but longer over the chine and thighs. Head fine, forehead straight; ears somewhat heavy, inclining horizontally; expression gentle but vivacious. Body somewhat lengthy, as in all good milking breeds, and well set on its legs. It is generally sought after as hornless, though when horns exist they are light and incline sharply to the rear of the head, curving round the ears." It is further described as one of the handsomest of goats. As a milker it is said to be remarkable, the specimens in the possession of the writer above quoted (which are only from 26in. to 28in. in height at the shoulders) giving easily 600 litres (130 gallons) during one lactation period.

The La Mancha has hair of medium length, and closely resembles the breed just described with the exception of its short, stubby ears. Several of this breed were exhibited at the Paris Goat Show in 1903, where I saw them, but with the exception of the peculiar ears there was nothing remarkable or very distinct about them.

CHAPTER VI

ORIENTAL AND AFRICAN GOATS

THE NUBIAN

THIS is a breed of which much is said by French writers, but with which we in England have had little to do as a milker. A few Nubians have certainly been brought at various times into this country, but they have either been males, or females that were not in-kid and giving little milk. The first known importation was that which took place in 1883. A pair named respectively Arabi Pasha and Aida were brought over by Mr. Paul Thomas from Paris, where they were said to have formed part of a herd descended from stock sent several years previously from Nubia for the Société Nationale d'Acclimatation. The second importation was in 1891, when from the same source another Nubian was obtained which was named Ali Baba. From what we know now it is doubtful if the first pair were absolutely pure, as they showed a good deal of the features of the Syrian or Mamber goat. Ali Baba was of a different type, and from descriptions and photographs I have received since from M. Crepin was in all probability a specimen of the Berber variety, a name, by the way, which coincides somewhat singularly with the name given to the animal itself.

About the year 1900 two female goats said to be of this kind were sent from the Soudan, where goats of this type are very plentiful, by H.R.H. the Duke of Connaught as a gift to Queen Victoria, and these were placed at the Prince Consort's Shaw Farm at Windsor. The late Mr. Tait, who had charge of this farm, desiring to breed from the animals, sent them to a three-parts Nubian male, then in my possession, by Sedgemere Chancellor, by whom they were served, but nothing came of the service. On the death of the Queen these goats were transferred, with other animals from Windsor, to the Zoological Gardens in Regent's Park. An application was subsequently made by the British Goat Society to the King for permission to take charge of, and attempt a second time to breed from, these animals, and consent was very graciously given by His Majesty. One goat went to Mr. Ravenscroft, of St. Albans, where it was successfully mated, but it unfortunately died during parturition, the kid, which was of enormous size, having to be destroyed in the hope of saving the life of the dam. The other and older goat was sent first to Mr. H. E. Hughes, of Broxbourne, and subsequently to Mr. Sam Woodiwiss, of Great Waltham, but, though repeatedly served by different males, she never brought forth kids, and eventually died of old age on the farm of the latter breeder.

The next importations were males only. One of these was obtained in 1903 by Mr. Sam Woodiwiss from the menagerie of Mr. Geo. Sanger, and named accordingly Sedgemere Sanger. This animal, which attracted considerable interest at the time, had all the characteristics of the Zareber as given below, and may therefore be regarded as a true specimen of that variety of Nubian. The other, Bricket Cross, was only nominally Nubian, having been imported from the Chitral by Mr. Cross, of Liverpool, and bought at a high figure by Mr. B. Ravenscroft, who used him largely, until his death in 1908, to supply fresh Eastern blood to his Anglo-Nubian herd.

This is the sum total of our English experience with the Nubian, and I must therefore refer the reader once again to French writers, who have devoted much space to it in their works. It should be remarked, however, that neither Crepin nor Huart du Plessis seems to have had much personal experience with the Nubian *as a milker*. Both get their information chiefly from Dr. Sacc, a naturalist, whose statements, however, as to the yields of certain specimens are of so improbable, not to say impossible, a character that one cannot give credence to them. Thus we read in "La Chèvre" that Dr. Sacc, who had studied and bred this kind of goat at Wesserling (Alsace), stated that one milker in his possession gave from 10 to 12 litres (8¾ to 11½ quarts !). This is the yield of a good Kerry cow and the average daily yield of an ordinary dairy Shorthorn. A goat would have to be as big as a fair-sized donkey to be physically capable of elaborating it. It is remarkable how people exaggerate on this point.

According to the writers above mentioned, the introduction of the Nubian breed into France seems to have been due to the gift of a young hippopotamus which some Eastern potentate made to Napoleon III, and which was accompanied by some Nubian goats to supply its daily needs. These found their way to the Jardin d'Acclimatation, where they came under the observation of the then Director, M. Geoffroy St. Hilaire, who was so astonished at their milking powers that he preferred not to mention the precise amount supplied for fear of being charged with exaggeration. The worthy doctor previously alluded to was evidently less reticent. M. Crepin had himself a small herd of Nubians, which were exhibited on two occasions in Paris, where they each time came under my observation, but I have never seen them in milk. Referring to these goats in his book, we read as follows: "The herd of Nubians that we possess come from the confines of Erythrea, where they were obtained only after unheard-of difficulties. Apart from those resulting from several days' travelling on the backs of camels, it was necessary to resort to all sorts of artifice to escape the watchfulness of the local authorities, who strictly interdict the exportation of the kind we were seeking. This breed is called the Zareber,[1] and responds with great exactitude to the description given by Huart du Plessis, Geoffroy Saint Hilaire, and Dr. Sacc." The description given is as follows: "A close silky coat, long pendulous ears, an elongated neck and body, limbs long and slender; udder generally spherical, and so long and heavy that the natives are in the habit of suspending it in a leather bag to keep it from dragging on the ground.[2] Head short; the facial line is strongly arched at the upper part, and then depressed towards the nostrils, which are somewhat flat. The lower lip projects very often beyond the upper, exposing the front teeth as in the bulldog. The eyes are large and almond-shaped, and the cheek-bones are promi-

[1] The word Zareber, or Zaraibi (Fr. Zaraïbe) means an enclosure or stockade.

[2] It is inconsistent for a "spherical" udder to be of this description.

nent. The colour of the coat varies, the colours which predominate being a mahogany-red, nut-brown, white-cream, black, and a kind of bluish-grey. Three or four colours are often observable on the same subject. These are disposed in so diversified and so peculiar a manner that at a distance the animal gives the idea, according to its colour at least, of a cow rather than a goat." The quantity of milk M. Crepin announces as the utmost he has got (a yield, however, he is quite satisfied with) is 3 litres (a little over 5 pints). He adds: "At the age of three or four years, when their growth is completed, they will give at least double."[1] This is somewhat speculative and rather improbable. It is worthy of remark that M. Sacc speaks of the Nubian as a tall animal. Now, those of M. Crepin that we were acquainted with were quite diminutive, and he himself refers to them in the same manner, giving their height as from 65 to 70 centimetres (26in. to 28in.). Those we saw were not more than 24in.

THE SYRIAN OR MAMBER GOAT

Goats abound largely in Palestine and Syria, and are often met with in herds of 500 to 1,000. There are several varieties, but that which is most in repute as a milker is the Mamber goat. This animal is remarkable for its long pendulous ears, which are sometimes twice the length of its head and wide in proportion, with a slight curl outwards at the extremities. The hair is also long, especially at the hindquarters, but short on the head, though a tuft of long hair generally projects from the forehead. The facial line is straight, though often somewhat arched in the male. The horns incline to the rear, with a curl downwards like those in mountain sheep, but in the male are longer and take a decidedly spiral form. The colour varies, but a large proportion of these goats are either black-and-white or quite black.

A choice collection of these goats, described as the Gar-a, a sub-variety of the Samar, was exhibited by M. Crepin at the Paris Goat Show in 1903.

The Mamber goat is said to be a good milker, giving in full

[1] By a strange coincidence, at the time these words were written a letter was received from M. Crepin announcing his being obliged to dispose of most of his goats in consequence of taking up his residence permanently in Paris.

profit from 3½ to 7 pints a day. This can be readily believed, but it is less easy to credit the statement of Dr. Ira Harris, the American Consul at Tripoli, who is reported by Thompson in his book to have given the yield of some specimens as 16 pints a day ! The milk of the Syrian goat is further said to be exceedingly rich in butter-fat and of delicious flavour. It is the breed which supplies the famous butter of Aleppo, which is largely consumed in various cities of Asia Minor. Hoffmann regards the Mamber goat as a separate breed from the Syrian, which he states is similar in many respects to the Cashmere, but this latter statement is altogether extraordinary and one which I personally cannot endorse. The goat has, in fact, more resemblance to the Nubian than the Cashmere.

Indian Goats

The varieties met with in the East Indies are chiefly of the lop-eared type. The best are what are known as the *Jumna Pari*, which are bred along the Jumna River, and common on the Bengal side. They closely resemble the Nubian, having, like these, long, wide, pendulous ears, a bow-shaped face, and being tall and leggy. The horns are upright and often spiral-shaped, though in some cases they are closely recumbent on the head. The hair is very short, being beautifully sleek and glossy, and generally of a red-tan and black colour. They are said by residents in the district to be splendid milkers, carrying immense udders with big teats.

Another variety largely kept in India is the *Surat* goat, which is generally white and lop-eared. Specimens of both kinds were at one time brought into this country, having been used on board P. & O. steamers to supply milk to the passengers.

Chinese Goats

Besides the Cashmere, described in a subsequent chapter, which is found in some parts of China, there is a common variety known as the *Shân Yâng* (signifying "hairy sheep") met with in that country. It is a large kind with very long hair and straight horns ; the skins of these animals are frequently sold in England for mats and hearthrugs. Most of them are black, a very favourite colour with the Chinese, though white and grey ones are also met with.

The flesh of the goat is more extensively eaten in China than that of any other animal excepting the pig.

The Boer Goat of Cape Colony

This is a breed that partakes strongly of the Nubian, from which it is no doubt descended, as it shows the same wide pendulous ears, and in a modified form the arched facial line with the protruding under-jaw. Schreiner, in his work on the Angora, states that when the first Europeans landed at the Cape, they found the natives in possession of a short-haired variety of goats, practically indigenous to the country, but no goats are recorded as being in the possession of the colonists till 1691. The South African goat of the present day, known as the Boer goat, is, however, greatly superior in size and appearance to the original stock, due to "improvements effected by repeated importations of high-class goats of the common varieties of Europe." Judging by the Boer goat of to-day, I should say, rather, to importations from the East, probably of Indian goats, which is another proof of the estimation in which this type of animal was held as a milker over two hundred years ago. Referring to this goat, the writer above mentioned states: "It is a large animal, with powerful legs, a grand carriage, a bold, free step, and wild, prominent eyes. The coat is short, smooth, and coarse, of almost any colour, or combination of colours, frequently being dappled." To show the extent to which these goats are kept in the colony, the same authority adds: "The Boer goats have in many parts of the Midlands and the East been ousted by their beautiful rivals the Angoras; but still their number, according to the 1891 census returns, amounted to 3,444,019, or about 250,000 in excess of the number of Angoras."

The females, or ewes as they are there called, are kept for the supply of milk, but the males are in most cases castrated, and either killed for their flesh—for they fatten readily—or used, as they are largely, as leaders to flocks of sheep, being trained to the work and understanding certain words of command. "Two 'kapaters' (the local name for these 'neuter' goats) have been seen to lead a flock of 2,000 sheep through a good-sized river in lots of 100 or more at a time." A herd of some fifteen of these Boer goats was imported in 1880 by the Duke of Wellington of that day, when President of the British Goat Society, and two males were presented to the institution as travelling stud-goats for the use of members. The Boer goat has played a most important part in building up the Angora goat industry in South Africa, having supplied the mothers of nearly all the Cape Angoras. "But for the fact," says Schreiner, "that there were several millions of Boer goats thoroughly accustomed to the country, to furnish innumerable ewes for grading-up purposes, the industry would still have been in its infancy." A good idea of the Boer goat may be obtained from the picture of Sedgemere Chancellor which would very well pass for one of this breed.

The Dwarf Goat of Guinea

A description of the various breeds of goats would not be complete without some reference to this liliputian variety of the genus *Capra*. It can only be regarded in the light of a curiosity, however, being of little use for milk, although an African explorer who met with this animal largely in the neighbourhood of Lake Tchad reports it as giving a very fair amount of milk for its size. These goats, singularly enough, have been somewhat largely imported into Europe. Several were on view some thirty years ago at the Zoological Gardens, Regent's Park, one of which was exhibited at the Crystal Palace Show in 1875. Mr. Sam Woodiwiss in his early days of goat-breeding kept several specimens of this dwarf goat at East Finchley, and I have met with them also in the Jardin des Plantes in Paris.

CHAPTER VII

THE ANGORA OR "MOHAIR" GOAT

THE Angora, known also as the Mohair goat, has been regarded by some naturalists as a distinct species, approaching more in its appearance and characteristics to the genus *Ovis*. The face resembles that of the sheep in both outline and expression, whilst the similitude is rendered still more striking by the thick fleecy covering which envelops the animal from the top of the head nearly down to the hocks (Fig. 2). This analogy extends also even to its milk and flesh, the former being, though less abundant than in other breeds, richer in nitrogenous elements, and containing, like sheep's milk, a larger proportion of casein; whilst the meat has the flavour and texture of mutton. Even the bleat of this breed resembles more that of the sheep than the goat, the cry being softer and more approaching the "baa" of the lamb. It is doubtless owing to this resemblance that Pallas regards the Angora breed as a cross between the goat and the sheep, though it is the general opinion of naturalists at the present day that it is descended from *Capra falconeri*, whilst the varieties described in previous chapters have their origin in *C. ægagrus*. In spite, however, of its resemblance to the sheep, there are many points which clearly denote that the Angora belongs to the genus *Capra* and not to the genus *Ovis*.[1] This is particularly exemplified in

[1] Mr. S. C. Cronwright Schreiner, in his book "The Angora Goat" (published under the auspices of the South African Angora Goat-breeders' Association), entirely disagrees with these statements, both as regards the Angora being a distinct species and as to the suggested resemblance between it and the sheep. In a footnote in the chapter "Descent of the Angora from the Wild Goat" we read: "The idea, however, of there being at least a close resemblance between the Angora goat and the sheep has been very widespread. Pegler remarks that the Angora approaches more in its appearance and character to the genus *Ovis* (sheep) than to the genus *Capra* (goat)," further

the male, which, besides exhibiting that capricious nature so peculiar to its species, possesses also a beard—an appendage of which the ram is entirely devoid—and, above all, gives forth that peculiar odour natural to the male goat of every breed, though in a far milder form in this than in any other. The scent, indeed, is not noticeable at all in the Angora buck except during the rutting season.

History of the Breed

Although Asia Minor is at the present time the home of the Angora, it has been established beyond doubt that the introduction of the breed into that country is of comparatively recent date,[1] it having been imported originally from the mountains of Thibet. This gives colour to the statement that both this and the Cashmere are the produce of the same wild goat (*Capra falconeri*). Indeed, it is only within the last twenty or thirty years that writers have come to regard them as distinct, for in spite of the great dissimilarity in the appearance of the two breeds, the fact that they are both wool-bearers has been sufficient to identify them under the general name "Cashmere or Angora" goats, in which manner they are constantly referred to in old works. An account of the introduction of the Angora into Asia Minor was published some years back in the *Field*, from which I give the following extract :

(quoting from the text above) : "Dr. Hayes says 'The cry is wholly different from that of the common goat, and resembles that of the sheep.' Even the 'Encyclopædia Britannica' falls into the same error. It says 'the face has a sheepish expression. . . . This variety of goat approaches nearest in its nature, form and habits to the sheep, even the voice having a strong resemblance.' Other authorities might be adduced expressing similar opinions. I have only to say that, having farmed Angoras for many years, having seen nearly all the best animals in South Africa (many of them being of unmixed descent from imported stock) and all those recently imported . . . I have never noticed any resemblance whatever, in cry, habits, or appearance, to the sheep ; nor have I heard of anyone else in South Africa having observed it. . . . The fact is, no such resemblance exists : the Angora is a perfect goat, in cry, habits, and appearance, only it is the most refined, highly-bred, and blood-like of all the various varieties of domesticated goats." It will thus be seen that opinions differ on this subject.

[1] The Mohair goat, spread over a large extent of Turkey to-day, is a product of, at most, the last eighty or ninety years, mainly of the last fifty or sixty years, and differs materially from the original Angora, which was confined in a remarkable degree to a small tract of country peculiarly adapted to it. (Schreiner on "The Angora Goat".)

"The first European record of the Mohair goat dates from the year 1541, when Busbek, the then Dutch Ambassador of Charles V at Constantinople, managed to secure a pair of these animals, and sent them as a curiosity to his Imperial master, with a strong recommendation that they should be introduced into Europe. Busbek explains that he was informed they had been recently introduced into Asia Minor from Armenia.

Fig. 2
ANGORA MALI GOAT

All later inquiries support this theory, for in districts where they have succeeded best the graziers assert that they came from the eastward. According to tradition they were first kept as household pets, and they still retain in a high degree the gentleness and tameness derived from this early method of domestication. The beauty and silkiness of their fleece attracted the admiration of the female members of the household, who quickly appreciated its value as a fibre for the private manufacture of articles of adornment in female attire, each family at that period keeping from five to ten goats for their especial use. All the extra fleeces not required for the purpose mentioned were used for stuffing beds. In this way this delicate and valuable animal was preserved before mohair came to have a commercial value. At the present time it is extinct in the regions where it derived its origin."

VARIETIES

"The most northerly point at which the Mohair goat thrives is Kastamboul," says Mr. Gavin Gatheral,[1] "a large and fertile province, but too near the moist winds of the Black Sea to reach its highest development, the fleece, though lustrous, being hard and coarse. Angora, the capital of the province of that name, produces five different varieties from as many districts, each equal in area to the largest English county. Yabanoon produces a heavy, lustrous fleece. Tchorba, a mohair so soft and fine that it falls to pieces as soon as shorn from the goat's back. Tchiboukooa is remarkable for its length and fineness. Ayash produces a white but lustreless fleece. The rams of the three first-named districts are undoubtedly thorough-bred, and though smaller in size than some other varieties, have all the points that a practical stock-breeder commends. . . . Jewar, or Near Town, is bright and showy, but full of what is technically termed 'stick,' or 'kempy' hair. Beybazar, or Princes Market, is so near Angora that the mohair it produces has no marked points of difference. The ram is larger in size, very heavy, and stands a sea voyage well. A few have recently been imported into Cape Colony and California, the result in both instances being highly satisfactory. Tchukess (or 'Circassian village') and Geredeh (or 'behind the mountain') are two districts where the Mohair goat has been introduced in comparatively recent times, and where, although stocked from other districts, they develop distinct characteristics owing to the difference of climate and elevation. The Geredeh ram is a large and powerful animal, covered with a fleece that seems almost black, so surcharged is it with grease; but when scoured the mohair is found to be second to none in quality and fineness. . . . The fleece of the Komah variety is reddish brown, and though this reduces its value as mohair, it is sought after for certain special manufactures. On the frontier of Armenia and Mesopotamia, and far to the eastward of this district, is a province called Van, which has hitherto supplied a great weight of inferior mohair, more resembling sheep's wool than goat hair."

[1] British Vice-Consul at Angora, in a paper read before the Royal Colonial Institute at Port Elizabeth.

Besides the difference in the quality and colour of the fleece, however, there is considerable diversity in other external features of the Angora. Thus the ears vary both in length and degree of pendulosity, the greater part being semi-pendulous, but in some more horizontally inclined. The horns, too, differ ; many are nearly perpendicular, with a slight spiral, others show a lateral growth with a distinct twist, whilst here and there they are met with growing back over the head with a slight inclination in an outward direction towards the extremities. These variations are due in great measure to crosses more or less remote with a common kind known as the Kurd race, which is distributed generally over Asia Minor. It is a black goat, rather larger than the white or Angora proper, the hair of which, although long and of a fleecy nature, is perfectly straight, without curl, being at the same time coarse in quality. Crosses with this variety are not infrequently made, the object being to gain hardiness, and the purity is restored, it is stated, from the third to the fifth generation.

In the spring of 1879 a valuable herd of thirty pure Angoras, procured from Asia Minor at a cost of over £1,000, were transhipped at the Victoria Docks on their way to Port Elizabeth, South Africa. Through the kindness of the importer, Mr. J. B. Evans, whom I shall have occasion to mention again presently, an opportunity was afforded me of inspecting these goats, besides being made acquainted with the correct points of the breed according to the judgment of the professional stock-breeder. These are as follow :

Points

Head fine, with the fleece growing well over the forehead ; ears wide, thin, and pendulous, not long and hanging down direct from the head, but standing out and then lopping over, more resembling the ears of a mastiff on the alert, and best described perhaps as semi-pendent. The horns are flat-shaped ; they should be set far apart on the head, and taper gradually towards the tips. Their direction depends on the sex. In the male they should incline first to the rear, with a slight twist outward, and the ends pointing upwards. In the female, on the contrary, they take a lateral direction, the spiral being more decided, and the extremities pointing downward. The chief feature, however, necessarily consists in the length, texture, and character of the fleece, which should be as free from anything like hair as possible. In the best specimens it is of a fine silky nature, growing in thickly-matted flakes near the skin, and then separating into long corkscrew-shaped ringlets, covering the whole animal as far as the hocks.

It is this material which is known commercially as mohair, and which is shorn off every spring for manufacture into fabrics. If not removed in this way it falls off naturally as the summer approaches. Besides this fleecy covering, however, there is an undergrowth of ordinary hair which is very short and lies close to the skin, being no doubt intended by Nature as an additional protection against cold when the animal sheds its outer coat. In common kinds the average weight of fleece in a herd, reckoning full-grown goats and kids, is about 2½lb., but in the best breeds it attains an average of 6lb. The greatest weight is taken from the rams, which shear as much as 10lb. and 12lb. each, those that are castrated, and between two and four years old, producing the finest quality. The length of the strands is about 8in.

Suitable Climatic Conditions

The Angora stands cold well, providing the climate is a perfectly dry one. In a communication made to Dr. Hayes,[1] Sir Samuel Wilson, whose experience in breeding Angoras under varying conditions is considerable, states that "a dry and moderately warm climate suits the Angora best; but altitude is of less importance. I have kept them in plains 500ft. above the sea, with an annual rainfall of 18in., and a warm climate with no snow and little frost, and they succeeded admirably. Also 1,800ft. above the sea, in a latitude colder by three degrees than the above, where the rainfall is 27in., and they did not do so well. Finally, at 1,200ft. elevation, on open bare plains, without snow in winter, or very rare, and they did fairly well; but it is a little too cold. The rainfall there is 25in."

[1] Author of "The Angora Goat: Its Origin, Culture, and Products."

INTRODUCTION INTO OTHER COUNTRIES

The great increase in the demand for mohair of late years has led to various attempts being made to breed the Angora in other countries besides Asia Minor. This has been done in the United States (Texas and California), Australia, and Cape Colony. In Texas the attempt has been made with some good results, as also in California, where pure stock was first imported from Asia Minor in 1861; in neither of these countries, however, can it be regarded generally as a commercial success. Regarding California, it is stated that though the appearance of the animal is prepossessing, and the wool commands a high price in the market, the yield is not sufficient to pay the interest on the cost and the expense of the keep.

In the United States the first importation was made in 1848, by Dr. J. B. Davis, who at the time was U.S. Consul in Turkey. Since then some hundreds have been introduced and bred in herds by skilled agriculturists, one of the most prominent being Colonel R. Peters, of Atalanta, who after an experience extending over many years regarded these animals as a valuable acquisition to the resources of the country. From this remark, and considering the known go-ahead, enterprising character of the Yankees, who use mohair in enormous quantities, one would have expected to see the Angora ere this as completely established in America as the sheep; but though Angora breeding is fast developing such is not yet the case.

In the Antipodes the first importation was made in 1853 by the Acclimatation Society of Australia, the goats being located in the Royal Park at Melbourne. These answered so well that in 1866 a further importation of ninety-three head was made, and the whole flourished and increased amazingly. In 1870, however, a large proportion of the herd was sold, and specimens distributed over the country, some being transferred to the care of Sir Samuel Wilson.

As far as commercial results are concerned, in no country has the introduction of the Mohair goat met with so much success as in Cape Colony, where the raw material is now produced in such large quantities as to form an important item in the resources of that colony. The following figures show the quantity and value of mohair exported from the colony, and the rapid increase of this industry :[1]

In 1857	870 lb.	value	£10
,, 1867	50,832 ,,	,,	£1,963
,, 1877	1,433,774 ,,	,,	£116,382
,, 1887	7,153,730 ,,	,,	£268,446
,, 1897	12,583,601 ,,	,,	£676,644

In 1927, from the Union of South Africa 10,681,290 lbs. of mohair was exported, and the goat population of the Union was over 7,500,000. The first attempt was made here in 1857, when a few pure-bred rams were introduced and crossed, for want of a sufficient number of pure ewes, with the indigenous Cape goat. These half-breeds produced a coarse kind of mohair, which, although fit for little else than stuffing saddles, proved sufficiently saleable to encourage the experiment on a larger scale, and a further importation from Asia Minor was accordingly undertaken. This second venture, consisting of six rams and one ewe, met with marked success ; the animals were offered for sale by auction, and realised high prices, one ram fetching £100 and another £117. Their progeny, however, soon became scattered far and wide, and, being crossed with native goats, without proper judgment, the purity of the breed greatly deteriorated. Later importations greatly improved matters, which was clearly evidenced by the higher prices obtained for the clips in the English market. This circumstance encouraged Mr. J. B. Evans, acting as manager to the Cape Stock Farming Company, to undertake a journey into Asia Minor, and to penetrate into the heart of the Angora breeding district, there to select the finest stock that could be obtained for money. The result of his purchase has already been alluded to ; but to show the superiority of these animals, and the high value that was set upon them on the arrival of twenty out of the original thirty head which he succeeded in landing at the Cape, I may mention that five of the rams realised, when sold by auction, an aggregate of £1,755, their respective prices being £240, £270, £395, £400 and £450.

[1] "The Cape now yields about one-half of the world's supply of mohair, or about the same quantity that Turkey yields" (Schreiner).

The goat which attained the last remarkable figure is the subject of the illustration on another page, and being drawn from a photograph of the animal taken in full fleece, just previous to its embarkation at Constantinople, may be regarded as a perfect type of the breed, though in some the fleece is still longer.

Experiments in England

The experiment has been tried both in this country and in France of breeding the Angora for its fleece, but it has never been carried on with any great success, neither the soil nor the climate being suitable; for though the animals have lived and even thrived, the quality of the mohair has deteriorated. One experiment in England was attempted by the then Duke of Wellington. In 1881 he imported from the Cape a herd of some half a dozen, which he placed in his park at Strathfield-saye, near Winchfield, Hants. For the first six months these animals seemed to thrive well, and a good clip was secured, but as the winter season came on, several died, and by the following spring the herd was very much reduced in number. This was doubtless owing to want of proper management, there being no one who understood their requirements. Later on, however, this difficulty was to a great extent remedied, and the goats again increased and multiplied, till at the Duke's death there were about five-and-twenty, including kids. From these animals a good quantity of fleece was collected, though the quality was coarser than that obtained from either Asia Minor or the Cape. It served, however, for manufacture into a coarse kind of cloth, being mixed with the hair from some llamas, also kept on the estate. Many of the clothes worn by the Duke were made of this mixture, and I myself possessed an overcoat cut from the same stuff. The Strathfieldsaye herd was in 1885 sold by auction, when I took the opportunity of securing two couples of young stock; these, during the six years I had them in my possession, did well, but there was no demand for the breed in this country, so I disposed of them.

My own experience, comparatively short though it has been, combined with the results I have seen obtained by others, as related above, leads me to the belief that the Angora goat, by proper housing and judicious management, will live and thrive in this country, though the climatic conditions are not favourable to the growth of fleece equal in quality to that imported. Nevertheless, I believe it would be good enough for many purposes to which inferior mohair from abroad is now applied.

Feeding and Treatment

As far as I have seen, I do not find that the Angora requires feeding or treatment different from that given to other goats. They do well stall-fed, if housed in a *dry* place, but though, like all the goat species, they are fond of a change to leaves of trees and brambles, they are well adapted to pasturage, and less mischievous and addicted to roaming than the common goat. Although, like the latter, they are said to object to rain, I have seen them in the park at Strathfieldsaye grazing in a herd by themselves under a steady downpour that would have caused any other kind of goat to seek shelter.

The Angora has seldom more than one kid at a birth, and is not a good milker, rarely giving more than a quart a day, sometimes less, though the milk seems more concentrated. At kidding-time they require some extra attention, it being often necessary to put the kid several times to the dam to get it to suck. If the weather be cold and the kid weakly, it will require to be brought into the house and placed before the fire, and a few drops of warm milk from the goat put into its mouth. Once it takes to the teat, however, it is able to look after itself. The young of the Angora grow more rapidly than those of the common goat, though the increased size which the former present comparatively with the latter at the same age is partly due to the fleece they carry giving them a larger appearance.

Shearing in this country should be done towards the end of April or the beginning of May, as early, in fact, as the weather will permit, otherwise the fleece falls naturally, coming off on everything that the goat rubs against, and thus being lost.

In order to give my readers the benefit of experience wider than my own on this particular branch of goat-keeping, I will quote from Dr. Hayes' book some remarks made by large Angora breeders in America :

"Colonel Peters states that one shepherd with a good dog can readily attend one thousand head of goats during the summer months. They may be kept with sheep to great advantage: that is, in the same range, for they do not graze with the sheep, being more active and rapid in their movements. Feeding on leaves, weeds, and briars, they prepare the land for sheep, and do not lessen the number of the latter already kept in a flock. Under this arrangement the cost of attendance is very trifling, for they require very little attention, except at the time of bringing forth kids; and their habit of returning home every evening is a very valuable trait.

"The bucks come in season about the 1st of August, and can be relied upon for good service until the month of February. Ewes come on heat during September. Young ewes with their first kids require to be kept in an enclosure, so as to prevent them deserting the kid."

CHAPTER VIII

THE CASHMERE OR "SHAWL" GOAT

THERE are several varieties of this goat; but the true Cashmere is the goat of Thibet, which is met with generally between the principal and secondary ranges of the Himalayas, near the regions of perpetual snow, some 15,000ft. above sea-level.

It has often been supposed that the Cashmere is identical with the Angora, but this is quite a mistake. Some naturalists, as already observed, regard it as descended from the same wild goat, *i.e.*, *Capra falconeri*, which may not be improbable, seeing that this animal frequented the mountains of Thibet, and that both domestic breeds are wool-bearers. Beyond this, however, the two goats have little in common, and even in regard to their fleece there is an important distinction, for whilst in the Angora it is the outer covering that is of a woolly nature and manufactured, in the Cashmere it is the undergrowth which is fleecy and utilised in commerce, the outer covering being hair-like in the common breeds, and comparatively valueless. In many respects, indeed, the Cashmere presents a much stronger resemblance to the common goat than to the Angora. It is a rather small race, with hair measuring from 4in. to 5in. in length, generally white, with sometimes reddish patches on the neck and shoulders. A few specimens of this breed were on view at the Zoological Gardens, Regent's Park, in 1877, and subsequently, having formed part of the collection of animals brought over from India by King Edward, then Prince of Wales. As these were presented by some Indian potentate, it may be reasonably presumed they were pure-bred and represented the true type, more especially as their appearance corresponded to that of others placed in the Gardens four years after. These goats were small-sized, with semi-pendent ears and twisted horns, inclining at their extremities to the rear. Schreiner states: "Before the Angora goat was introduced into the Cape efforts had been made to import and breed the Cashmere. . . . The Angora goats imported into the Cape were for a long time known as Cashmeres."

The points are a short, delicately-shaped head, thin ears, small bones, and a long heavy coat; the longer the hair, the more abundant the fine undergrowth it covers. This undergrowth consists of a beautifully soft downy wool, more fleecy in its nature than mohair, very short, and of a white or greyish-white colour. It grows at the roots of the long hair, appearing in the autumn, and being shed in the spring, when it is collected by a very careful combing process, occupying from eight to ten days. This is a work of great patience, as the undergrowth requires to be carefully separated from the hair that becomes detached in the operation and accumulates on the comb. The quantity obtained even in the best specimens only amounts to about half a pound, whilst in inferior stock it varies from 4oz. to 6oz. It ranges in price from 5s. to 6s. per pound. This is the material from which the far-famed "Cashmere shawls" are made in India, where at one time 16,000 looms were in constant work in their manufacture.

As may be supposed from the great elevation at which the Cashmere goats pasture on their native mountains, they can endure a considerable amount of cold; they are, in fact, a particularly hardy breed so long as they are kept in a dry climate; but a damp locality kills them. Attempts have been made in England and France, as also at the Cape and in the United States of America, to acclimatise the Cashmere, but with indifferent results. As early as 1819 Baron Fernaux and M. Joubert introduced a herd of nearly 400 head into France, some of which under favourable circumstances are stated to have thrived well enough, but their fleeces degenerated in quality, the amount of material obtained from each individual being at the same time exceedingly small.

Introduction into England

A portion of the above herd was purchased by Mr. C. T. Tower, an English gentleman who happened to be in Paris at the time they arrived, and who removed them to his park at Weald Hall, Essex. Here they flourished sufficiently well to enable their owner to have a shawl made from their fleeces which were pronounced by competent judges to be of very fair quality. A herd of goats descended from this importation still roamed at large in the same park in 1909, in the possession of Mr. C. J. H. Tower, grandson of the original importer.

The Windsor Herd

About the year 1828 a pair of goats from the herd mentioned above were presented by the owner to King George IV and placed in Windsor Park, where they rapidly increased, developing in course of time into the large herd which roams the Park at the present day. Here the experiment which had been tried by the proprietor of Weald Hall was repeated, under the auspices of the Prince Consort, whose well-known interest in all that concerned the industries of this country led him to enquire further into the practicability of utilising the home-grown produce for commercial purposes.

Referring to this experiment, Sir Samuel Wilson, in his work on the Angora goat, writes: "Desiring to have some fabrics manufactured from the fleece, a quantity of the wool and hair, as it was shorn from the goats, was sent to a large manufacturer. The separation of the wool from the hair being at that time—from the imperfection of the machinery in use—a very difficult operation, a great number of ladies assisted the manufacturer by taking small portions of the fleece and picking by hand the wool from the hair. Such was the enthusiasm caused by the experiment that over a thousand persons of all grades and conditions were employed in the work. Each person so employed received as remuneration an elegantly engraved certificate stating that the holder had assisted in bringing to a successful issue the experiment of His Royal Highness in the manufacture of Cashmere goats' wool. Some brocades and two beautiful shawls were produced by Messrs. Haley, the manufacturers, and the hair was made into a coarse fabric which was shown in contrast with the finer textures."

In 1889 a consignment of Cashmere goats was sent to Queen Victoria from India, to provide fresh blood to the herd in Windsor Park, which by that time had considerably deteriorated through in-breeding. This was the year when the Jubilee Show of the Royal Agricultural Society of England was held in Windsor Great Park, on which occasion prizes were awarded for goats. Here the British Goat Society had a stall, and amongst the exhibits were an Indian Cashmere shawl, lent for the purpose by Her Majesty Queen Victoria, and the shawl referred to above, lent also to the Society by Mr. C. J. H. Tower, one of its Life Members. The latter shawl was submitted to the Queen's inspection and much admired.

The Stow Park Herd

The Duke of Buckingham and Chandos, in 1863, had a herd of Cashmeres in Stow Park ; these he received from his uncle, the Marquis of Breadalbane, of Taymouth Castle. Exchanges were at various times made between the two herds for change of blood.

CHAPTER IX

CROSS-BRED GOATS

THERE is such a vast number of cross-bred goats that it is advisable to offer some remarks upon the crosses that may most advantageously be made. From my own experience, excepting the Toggenburg, I do not consider that for practical purposes a pure specimen is always the best. Of course, when for any special object it is required that a variety should be kept pure, as in breeding Angoras for their mohair, for instance, it is a different matter ; but as far as milk goes, a cross-bred goat, so long as it possesses a good milking pedigree, is quite equal to one of pure blood. I have rarely, however, found it satisfactory to breed a short-haired with a long-haired goat, and therefore do not recommend an English and Irish cross. The produce is generally a nondescript kind of animal, with long hair on the ridge of the back and on the hind legs, and sometimes a good deal under the belly. The horns are usually long and coarse, and more branching than the Irish, though the general appearance of the latter predominates.

The cross between the English and the Nubian has already been described (see pages 28 to 30), but other crosses have been tried in England, such as that between the English and Toggenburg, the English and Saanen, and the Toggenburg and Alpine, many of which have turned out to be exceptionally good. There is no doubt that by a combination of the Nubian or Anglo-Nubian and the Swiss breeds like the Toggenburg, Alpine, or Saanen handsome animals and splendid milkers are to be obtained. The poor quality milk, as regards butter-fat—which is the one failing of Swiss goats as milkers—is compensated for by the Eastern blood, the milk of Anglo-Nubians being much richer. In all such crosses I should prefer to use the Toggenburg or Saanen as the male, these breeds being the purer-bred animals and stamping their features in a more pronounced manner on their offspring than the Anglo-Nubian. As examples of successful cross-bred goats coming under the above description may be mentioned Miss E. Mortimer's Broxbourne Topsy, a champion winner, and Miss E. M. Pope's Broxbourne Dorothy, first prize in its class and the reserve number in the milking competition at a Tunbridge Wells Show, the latter an Anglo-Nubian-Saanen, but showing chiefly the colour and characteristics of the Saanen. Both these goats were bred by Mr. H. E. Hughes.

The crossing of the Nubian with the common short-haired goat has been practised in France with the same degree of success as in England, though on a much smaller scale. In M. Crepin's book "La Chèvre" two goats are depicted, each out of a half-bred Nubian by a pure Alpine, and the author reports that they excel as milkers both of the breeds from which they are the produce. The appearance of these animals fully justifies the statement. There can be no doubt that the chief advantage gained by this crossing of breeds is the introduction of absolutely new blood, and the effect of this is very quickly observable in improved stamina and a more abundant milk-flow. In England, where goat-breeders have so few opportunities of importing stock from abroad, and where there is always a run on stock that are successful as prize-winners (especially on milking-prize winners), there is generally a certain relationship between the best strains, so that when a cross-out can be obtained with some good milkers, such as the Toggenburg or Alpine, the result is generally a marked success.

Crossing with the Angora Goat

I cannot help thinking, and I believe I am not alone in this idea, that it would be quite possible, by carefully crossing some of the best milking strains of the common goat with the Angora, to produce an animal that would embody in itself the qualities of both breeds, and whilst giving a plentiful supply of milk would yield mohair of sufficiently good quality to be marketable, thus making the cross doubly profitable. Such a cross, however, would be of little use in England. Regarding this question, the following remarks, contained in a letter from Colonel J. W. Watts to Colonel R. Peters, which I extract from Dr. Hayes' book on the Angora goat, are extremely interesting:

"I began with the milk goat, had two varieties—the long, rough-horned Maltese, and the short-haired South American. About the year 1872 I came in possession of three Angoras—one billy and two nannies. . . . I bred this billy to my milk goats, the result being a beautiful lot of large, well-formed half breeds, showing no fleece. He was bred to his kids the next fall; this produced a pretty animal with considerable fleece, which grew to be about 3in. in length. Those descended from the short-haired goat had a much more uniform fleece than those from the long-haired goat, and some of those retained their long straight hairs through the fourth and fifth crosses. The length of fibre is about 3in. on the second cross, and continues to increase in length on each cross until five crosses are made, when it is about equal in appearance and value to the thorough-breds. Whilst the thorough-bred has rarely more than one kid at a birth, the grade (cross) usually has twins, which they raise well, especially those graded from the milk goat, as they retain their milking properties for several generations. I had a half Angora and half Maltese ewe that gave 4 quarts of milk per day for months, and after half of her udder was destroyed she raised two kids each season until her death, usually the largest on the farm."

Many years ago a pair of Angora goats were sent over from Australia as a present to the Baroness Burdett-Coutts, and located at Highgate. These were crossed with some half-bred English and Nubian goats on the estate, and produced some very handsome stock, partaking strongly of the Mohair breed.

Can the Goat be Crossed with the Sheep?

This seems a moot question, as authorities differ on the point. Colonel R. Peters, a contributor to the *American Agriculturist,* who has had great experience with Angora goats, writes: "Prior to the year 1860 I tried many experiments, in hopes of obtaining a cross between the goat and sheep, and failed in every instance. Extensive correspondence with other breeders has convinced me that the cross cannot be obtained. Dr. J. Bachman, the celebrated naturalist of Charleston, S.C., who was in correspondence with the most distinguished naturalists of Europe, informed me that he had no faith in the theory, and did not believe the cross obtainable." On the other hand, Sir Samuel Wilson states that the cross between the goat and the Merino sheep has been tried, and the progeny was not unfertile, though it reproduced with difficulty. He mentions that "Professor Cretzchmar, a learned naturalist residing near Frankfort-on-the-Main, experimented on twelve Merino ewes and an Angora buck, with the hope of establishing a breed of animals intermediate between the two, but it was not till the third season that the experiment succeeded, and the progeny so closely resembled the Merinos that little difference could be noticed in their external characters."

Quite recently Mr. Sam Woodiwiss kept an Angora she-goat with a Suffolk ram during the breeding season, but nothing resulted therefrom.

In reference to "Supposed Goat-Sheep Hybrids," the following appeared in the *Field* under that title on 11th February, 1899, over the signature of W. B. Tegetmeier, the well-known naturalist: "The subject of supposed goat and sheep hybrids appears to be exciting some attention, and I have received several letters respecting it. Professor Newton has referred me to some interesting experiments in France, an account of which I hope speedily to obtain. Mr. C. Tindall, of Wainfleet, has sent me a communication informing me that an animal was born last week at Mr. Elsey's farm near Horncastle. He tells me that he saw the supposed hybrid, and could hardly think of it as anything else but a cross, which was the opinion of everyone who saw it. . . . The case is interesting, but it cannot be regarded as a definite one, inasmuch as there is no proof that the male parent of the animal was not a ram. Had the ewes been running with the goat alone the evidence would have then been definite and satisfactory; but we have seen in Professor Cossar Ewart's experiments that he failed to obtain any hybrid under those conditions."

CHAPTER X

SELECTING A MILCH GOAT

Points of a Good Milker

There are certain points by which a good milch goat may generally be recognised, and which may here be mentioned as a guide to the purchaser, in case he (or she) may not have experience with these animals. A good milker has a rather long but neat head, broad at the forehead, and tapering towards the muzzle, with horns (if any) fine—that is to say, thin—small, and tapering. A goat entirely devoid of horns is, in the estimation of most people, preferable, but their presence or absence has no bearing on the question of milk. The eye should be large and bright, and the expression of the face thoroughly feminine. This is an important point, though many people might not suppose it, a thick-headed, "billy-faced" animal, with large coarse horns and a masculine appearance, being rarely good for milking. The principal features, however, consist in the shape of the body, and in the udder and teats. Always look for a goat with a long body and a large deep frame, the ribs being well rounded, so that there is plenty of room for a big stomach; a heavy milker is always wedge-shaped—that is to say, it is much deeper at the hindquarters than at the chest. A broad chest is an indication of a tendency to accumulate meat and fat rather than milk, though it is also a sign of a good constitution. I never, however, regard a narrow chest in a milch goat as a defect, providing that it makes up for this by being wide behind. The best milkers are generally narrow-chested, with long thin necks, meagre bodies, and protruding hip-bones. Thinness is no drawback if the animal is a good feeder. The food must be transformed into something, and if a goat eats largely without getting fat, supposing the creature to be in health and milking, it must go to supply the mammary gland. Another point is the skin; this should be loose and supple, with hair rather soft and fine in quality.

As regards the udder, which is so often deceptive in appearance, let no one be attracted merely by the *size* of the bag independent of other considerations. A goat may have an immense udder and yet give a comparatively small yield, for the simple reason that it is composed mostly of flesh instead of milk. It should not only be large, but thin in substance and soft and elastic to the touch. When quite full it will be greatly distended, but after milking it should shrink up to a very much smaller size. Thus a goat which, when in profit, has a large udder, when she becomes dry may exhibit very little.

The teats should be situated fairly well apart from each other, and point forwards, the nicest being those that are long and tapering and of a size easily grasped in the palm. The udder should by preference be round rather than long and narrow, though it must be admitted that many heavy milkers possess bags of the latter shape.

Colour is a matter of taste. It has, in my opinion, no influence as regards the yield of milk; black goats are often considered the best, but I cannot say that I have found these preferable to any other.

How to Tell the Age of a Goat

Another important consideration when choosing a goat is the age.

The best age is about three years, after it has just borne its second kids, as the amount of milk given on the first occasion is often comparatively small. If a goat with her second litter has only just arrived at her second year, she must have had her first kids at too early an age, and before her growth was sufficiently developed, so that she would be stunted in size in consequence.

A fairly accurate estimate of the age of a goat may be obtained by looking into its mouth and examining the teeth; like sheep and cows, these animals have no incisors in the upper jaw, but only in the lower, it being by these that the age is to be ascertained. The mouth of a goat aged one year contains its full complement of teeth, thirty-two in number—namely, six molars on either side of each jaw, and eight incisors, or front teeth, in the lower jaw only. These are of small size, and rather pointed. In the second year (generally about the second month) the two centre ones fall, and are replaced by two new ones, easily distinguishable by their size, being considerably larger than the other six incisors. In the third year two more small teeth, one on each side of those already changed, are replaced, so that at that age there are four large incisors in the centre and two small ones at each end. In the fourth year the large teeth increase to six in number, and only two small ones, one at each end, remain. Finally, when the goat reaches her fifth year, these in turn fall, and are replaced, and she has then what is commonly called a "full mouth." After that time the means of ascertaining the age are less sure, the only plan then being to examine all the teeth generally, but particularly the molars or grinders; the more these are worn the older the animal may be presumed to be. At seven or eight years the front teeth begin to wear down, and if much used in barking trees they break or fall out without being replaced, so that a goat with one or more incisors missing, the rest being worn and broken, may be fairly assumed to be old and nearly useless, and should therefore be rejected.

I should state, however, that although these rules apply generally for distinguishing the age of a goat, there are so many exceptions that they cannot be always implicitly relied on. The precise age at which the teeth are changed varies greatly according to the conditions under which the animal has been reared, whilst the amount of wear and tear they have undergone after the "full mouth" stage is reached depends upon the usage they have been subjected to. If a kid has been forced by high feeding, its teeth are changed earlier than if reared less artificially, and sometimes even when the first pair fall after a year old the second pair will follow within a few months, the rest being rapidly changed in succession. Instances of this kind are clearly shown in the following table, which gives records of careful observations on this point:

GOAT No. 1			GOAT No. 2			GOAT No. 3		
No. of large Incisors	Age at which they appeared		No. of large Incisors	Age at which they appeared		No. of large Incisors	Age at which they appeared	
	Years	Months		Years	Months		Years	Months
2	1	2	2	1	2	2	1	1
3	—	—	3	1	7	3	1	10
4	1	11	4	1	8	4	1	1
5	2	3	5	2	—	5	—	—
6	2	5½	6	2	—	6	2	3
7	—	—	7	2	3	7	—	—

Goats that have done much browsing are sure to have their front teeth more worn down than those which have been stall-fed, the latter using their molars more than their incisors.

Indications of Health

Another matter which should not be overlooked in selecting a goat is its health ; nothing is more disheartening than when buying an animal of any kind which you imagine to be in perfect condition to find it in a dying state perhaps a day or so after. I can call to mind more than one occurrence of this character, and therefore as there are always unscrupulous persons ready to take advantage of the uninitiated, it may be useful to many of my readers to know how to detect a goat in health or sickness. When ill, these animals have a mournful, dejected appearance, their breath is offensive, and their gums of a pale colour ; their appetite is bad, and they lose generally that vivacity and sprightliness of manner which are so characteristic of them when well. When in good health, on the contrary, they carry their heads erect, their eyes are bright and sparkling, with a cheerful expression, their nose dry and nostrils moist, their breath is sweet, and their mouth and gums a bright red. Another sign by which their condition may be known is the vein of the eye. This may be examined by raising the upper lid with one thumb and pulling down the lower lid with the other. If the vein in the corner of the eye be a bright red colour, the goat may be considered to be in good health.

How and Where to Buy Goats

The great thing towards success in goat-keeping is to make a good start: that is, begin with an animal that will give satisfaction. Each year sees a fresh batch of recruits in the ranks of goat-keepers, some of whom, by going to work in a careful, methodical manner, achieve success, and become enthusiastic on the subject, whilst others, after the first season, give it up as troublesome and unprofitable. This frequently depends on the first goat bought. There was a time, before the British Goat Society came into being, when intending goat-keepers had to depend on general dealers to get them a goat, and the animal supplied to them was frequently obtained to order from a local market or country fair, the chances of such purchases turning out well being very problematical. With the British Goat Society and its present system of publishing its "Monthly Journal" all this is changed. There is always amongst its subscribers a certain number who have some goats to dispose of, and it would not answer their purpose to supply an animal that was not all it was stated to be. The Society used to issue a Sale Register Circular, but since the publication of the "Monthly Journal" this has been discontinued. The demand at the present time for good milking goats is so considerable, however, that to buy a really good milker with a Herd Book pedigree means a rather large outlay. Then there is always the chance of picking up a bargain through the private advertisement columns of *The Exchange and Mart*. In this journal articles and correspondence on goat subjects are occasionally published, and a number of goat advertisements naturally follow. By the "Deposit System" the transaction can be made through the management, in which case the purchase money is not paid until the deal is reported as satisfactory.

Caution to be Exercised in Buying

It is a singular fact that now, as thirty years ago, whenever a goat is offered for sale it is generally said to be giving, or to have given, when in milk, "two quarts a day." One might suppose this was the only measurement of a goat's yield known.

But measurement is always unsatisfactory. According to how a goat is milked, the animal may or may not appear to yield in two milkings the above quantity. A standard quart measure may be used and filled to the brim—*with froth*—before the eyes of an intended buyer, who, in his inexperience, will suppose he has there a quart of *milk ;* but, after standing some time, when the froth has settled, three-quarters of a pint will be nearer the mark. Now a gallon of milk weighs 10lb., therefore a pint should turn the scale at 20oz., supposing the vessel holding the milk to have been counterpoised by a corresponding weight before the milk was introduced. This is the only proper way to ensure satisfaction in the question of yield. To see a goat milked before the eyes of the purchaser once is not enough, as it may have been "stocked" for twenty-four hours, and then the next milking will be reduced by half. A goat sent on trial should stay for a couple of days at least, but preferably three days, in order to let it get accustomed to its new surroundings, and even then one must not expect to see in a strange place the same yield forthcoming that was given by the animal in its own home. A goat is a highly nervous and very capricious animal, and its yield is readily influenced by change of surroundings, change of food, and especially change of milker. For, like a cow, it will not give down its milk properly at first with strange hands tugging at its teats.

All these circumstances must be taken into account when a goat is sent on approval to a new home, and it is therefore always more satisfactory to see the goat under selection at the owner's house and milked by the usual attendant, when all disputes can be thus avoided.

Selecting a Breed

I devote a paragraph to this matter because one is so often asked about it by people who have no experience in goats, and who think they have the same choice as regards varieties as if they were dealing with poultry or dogs. There is, as a matter of fact, very little scope for the goat fancier in the way of selection of breeds. It may be all very well to point to the foreign varieties mentioned in the earlier chapters; such goats, with a few exceptions, are not for the *British* goat-keeper. He can take only what these islands contain. A time may come when facilities for importation will be greater than now, but dealing with the present, the would-be goatkeeper has many restrictions. New British breeds are now being established which will make a greater selection possible, but hitherto, the only pure breeds the United Kingdom could offer were Irish goats, or the common English goats—and purity here is questionable—Toggenburgs, and Anglo-Nubians. The first two only are the genuine *vacca pauperis*, the third are the goats for the rich, and the last those for the average goat-keeper who can afford a five-pound note for a fairly good animal. The Irish goat not being recommended owing to its long hair, the choice really rests between the short-haired, prick-eared English goat, the Toggenburg (including crosses with these and the Alpine known as "Swiss"), and the Anglo-Nubian. The first of these are fairly plentiful and cheap, and answer very well to begin with. The Swiss pay best to keep because the milk is generally more abundant, and the demand for stock being great the kids fetch high prices. As for the Anglo-Nubian, if obtained from *a good milking strain*, they have advantages which are likely to appeal to the novice. Many people prefer the handsome lop ears and close shiny coat which make the kids such perfect little pictures. After all, purity has ever been a small matter in goat-keeping, and a person who can secure a strain of goats that breeds milking-prize winners will soon get a demand for stock at fancy prices, whether they be English, Toggenburgs, British Alpines, Anglo-Nubians, Anglo-Saanens, or other crosses. If we except the Toggenburgs nobody bothers much about pure blood, but everyone is keen to get good milkers of the popular type, which is a short-haired, hornless goat with a long body, slender neck, neat head having either pricked or drooping ears, and above all a big udder. Such goats win their prizes at shows and fetch their £15 and £20, and no one cares whether they are pure-bred or absolute mongrels, so long as they have won their laurels in a milking competition.

Advice at Starting

Considerable advantage may be derived by keeping two goats instead of one, the amount of time and trouble required being scarcely any greater, whereas the profit derived is more than double. The best plan is to begin by purchasing one that has just kidded, and when that is becoming dry, to procure another just about to kid; by this means a good supply of milk will always be kept up.

As some goats are spiteful and bad-tempered, and consequently troublesome, especially at first, to milk, if those who are to have the care of them are women or children, it would be best to begin by purchasing a female kid about six months old, always supposing, of course, that you are not in immediate want of milk and can afford to wait a year before she becomes productive; by doing this you get an animal which will soon, by kindness and good keeping, become as tame and docile as a lamb, being almost as much attached to its master or mistress as a dog. In such a case it is more than ever necessary to buy from someone on whom you can rely, it being essential to ascertain the respective qualities of the parents.

Prices of Goats

In the earlier editions of this work I gave the prices at which goats could or should be obtainable, but experience has shown me that such information is bound to be unreliable. The prices at which goats could be procured twenty-five years ago would be of little use now, whilst different breeders have different prices, and the more a certain stock is in demand, the higher are the prices asked for specimens.

In reference to the prices of goats in-milk, it should be observed that these run much lower in spring than in autumn, as in the former season fresh-kidded goats are fairly plentiful, whilst in autumn they are exceedingly scarce and prices rule high accordingly. Big prices are not infrequently asked on account of a goat being in the Herd Book, but here it is well to remark that the mere fact of a goat's parentage being recorded in the Herd Book does not enhance its value unless such parents happen to be prize-winners, and especially milking-prize winners.

Advice to Sellers

I may here offer a word of advice to those having goats to sell. People often think because they hear of fancy prices being obtained by breeders of note that they can command the same for their animals, and are surprised that such goats do not sell when so advertised. Now, there are always plenty of persons ready to give £5 or £6 for a milking goat, but very few who care to pay much more. Breeders who get their £10 and £20 have generally spent large sums not only in the development of their herd, but in acquiring a name through winning prizes, and it often happens that a goat which would not realise £5 if advertised by an ordinary goat-keeper will fetch £10 in the hands of a well-known breeder and exhibitor.

A fair way of estimating the value of ordinary non-Show goats with a milk yield of less than a gallon is to allow £1 for every lb. of milk given by it daily when in full milk. The value of Show goats is calculated differently, as often a second-rate milker may win a number of inspection prizes and easily pay for its keep.

CHAPTER XI

THE GOAT-HOUSE

THOSE contemplating keeping goats will find the interest and pleasure in their management doubly increased if they have a clean and comfortable place in which to attend to the animals. Persons having an empty stable need, of course, no further accommodation, and only require to make the necessary alteration in the position of the manger or other arrangement for feeding to suit the height of the animals, and the thing is done. A horse's manger in an unused stable is best boarded over completely, when it serves very well as a sleeping-place for a goat.

There are few residences in the country that have not some kind of outbuilding which, with a little contriving, could be made into a comfortable goat-house of a more or less substantial character; but to those who have not such advantages, and are obliged to construct one for themselves, the following hints will doubtless prove serviceable.

The cheapest form of building will be a "lean-to," advantage being taken of the angle of a wall against which to erect it, if such presents itself; by so doing extra shelter and warmth, besides a saving of material, will be effected. The dimensions I am about to give will suffice for two goats, the cost of making being little more than for one, and I have already stated that advantage is gained by keeping a couple. Even should a single animal be the intended inmate, the increased space afforded will be found advantageous should any addition to the stock take place, as, for instance, when the kids make their appearance. The building should be 5ft. wide by 7ft. long, the length being divided as follows: 1ft. for the width of the manger, 4ft. for the length of the goat, and 2ft. for a passage to contain the few implements required for use in the stable. The height nearest the wall should be 7ft. 9in., the opposite side being 5ft. 9in., which gives a slope of 2ft. to the roof; this may be either tiled, or made with wood over which is laid a covering of felt. The latter plan is certainly the warmer as well as the cheaper of the two, besides being lasting, for good felt properly tarred when laid down, and retarred every second or third year, will stand nearly twenty years.

In constructing a building that is intended for the habitation of animals, great attention must be paid to one important point, viz., ventilation, without due regard to which no creature, however hardy, is free from the attacks of disease in some form or other. Means must therefore be provided for the escape of the heated foul air, admitting at the same time a corresponding amount of pure air to take its place. Care must be taken, however, to avoid any direct draught upon the goats, which would have an injurious effect, for, although fairly hardy as a rule, they are very sensitive to cold. Various means may be adopted for ventilating the house, such as the use of air bricks or louvre boards; but in a building of the nature and size I have described, nothing will be found easier to fix, cheaper, or better than pieces of perforated zinc, which should be placed in the highest part of the stable, just under the roof, another piece being inserted at the lower part, a few inches from the floor, just below the manger, thus causing an almost imperceptible current of fresh air to be admitted. "Hit and Miss" ventilators may, however, be used where expense is no object.

A good-sized door about 2½ft. wide running up to the roof will be required. The best are those which are known as half-doors, so that in fine weather the top door can be left open when the other is shut. A window is also necessary to admit light and air. One that opens by turning on a pivot in the centre will be found as cheap as any in construction, besides being easily opened and shut.

For the floor the best material is concrete; it is cheaper than bricks, and, being of a hard and smooth surface, is easier to clean. Care must be taken when laying it down to allow full time for the cement to set before it is trodden on; about twelve hours is sufficient for this as a rule, but in a damp place it will take two or three days before it becomes thoroughly dry and hard throughout. The floor must be raised about 2½in. from the level of the ground outside, to prevent the wet from entering under the door; it should also slope gradually from the manger to the entrance: this will be found of great assistance when flushing it with water, which requires to be done occasionally in summer. A small quantity of some disinfectant mixed with the water for flushing is valuable in deodorising and purifying the place.

If only one goat is kept, it may be allowed to run loose in its house, and a good plan then is to fix a bench against the wall about two or three feet from the ground. It will prefer this to lie on to any quantity of straw or other bedding placed for it upon the floor, and it is indeed healthier, especially if the floor is of a damp nature. Instead of providing a manger or trough to receive its food, it is better to have a plain board, 11in. wide and 1in. thick, in which two circular holes are cut sufficiently large to take a small galvanised iron pail, let down to within a couple of inches of the level of the board, which should be from 20in. to 24in. from the floor, according to the height of the goat, and supported on wood or iron brackets fixed to the wall. These pails, one of which is for water and the other for food, are much better than a manger, wherein refuse food accumulates and turns sour; for the pails can be readily removed to be filled, and any uneaten contents that may remain tipped out. At the same time, from the secure manner in which they are placed, the goat cannot upset them or waste the food by nosing about for the bits it most fancies, and turning out the rest.

When there are several goats, and especially when a he-goat is amongst them, it is most important that they should be all securely fastened up, and it is still better to provide a little stall for each, by erecting partitions at intervals along the feeding-board. These partitions need not, however, extend more than half the length of the animals' bodies, being merely required to prevent their fighting and interfering with each other whilst feeding. When this is not done, even supposing they are separated too far to do any actual damage, the attempts that are made by the stronger and more ill-tempered ones to interfere with their weaker and more nervous neighbours prevent the latter from feeding in peace. I much prefer in this case to place the stalls and feeding-boards away from the wall, so that food may be given them from the front. This I find greatly facilitates the operation.

GOAT-STALLS

It may, perhaps, be interesting if I give a description of the stalls in my own goat-house, so that those who may desire to imitate my plan and be their own carpenters can do so.

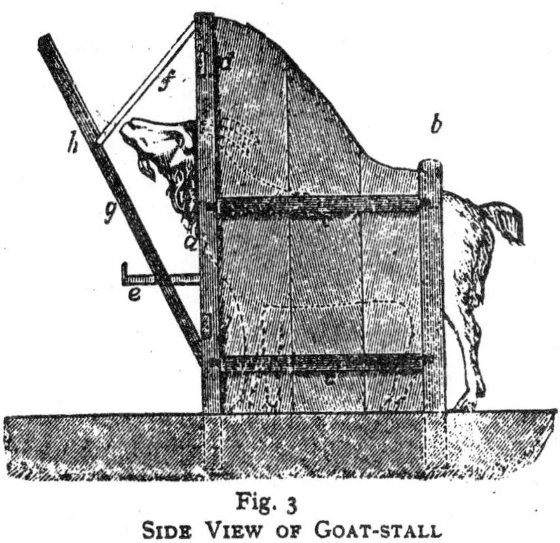

Fig. 3
SIDE VIEW OF GOAT-STALL

These are of various widths, to take different-sized goats, but the dimensions I would recommend generally are 2ft. for small she-goats and 2ft. 3in. for large full-grown animals. If the stalls are wider the goats are able to turn round, which is very objectionable, as they soil the front and sides. Besides this, the extra space enables them to make use of their horns against the divisions, which they will do very freely if they happen to have a neighbour they dislike, or if a new goat be placed next them. I will give in detail a description of one stall, whereby any number can be made that are required, it being merely a question of space and timber. The accompanying illustrations (Figs. 3 and 4) will assist the reader to understand my directions: Procure some quartering, 3in. by 2½in., and cut two lengths of 4ft. 6in., two of 3ft., and four of 2ft. The first are for the front posts (*a*), the second for the rear posts (*b*), and the third for cross rails (*c*), connecting the two posts together, which can best be accomplished by cutting mortices in the posts at distances of 1ft. 6in. and 3ft. respectively from one end, and making tenons at the ends of the rails to fit them. When thus connected they form the framework of the two divisions, which only require boarding up to be complete. Matched boards are the best to use, as they fit closer and look better; they should be 1in. thick, and shaped at the top after the fashion of the stalls in ordinary stables, as shown in the sketches. To make a proper job the cross rails should be rather narrower than the posts, so that when the boards are nailed on they come flush with them. When the partitions are erected the posts should be let into the ground about 6in. to give them strength and firmness. Two pieces of quartering are then nailed horizontally along the front of the stalls, either upon them as they are, or, what is better, let in by cutting a piece out of the post and a corresponding portion out of the rail to fit, known as a "half-lap," so that they do not project beyond the posts. These rails (shown in section at *d d*, Fig. 3) are placed one 15in. and the other 3ft. 6in. from the ground. If, as is likely to be the case, the end stall is against the wall, the cross rails should be let into the latter by knocking out a couple of bricks; this will make the partitions firmer. I find it best to keep the buck in this end stall, as he requires an extra strong compartment. The front of the stalls can then be boarded up in the same way as the sides, but leaving a space of 9in. for the head of the goat to pass when feeding. The boards must be nailed from the inside, otherwise an obstreperous goat, especially if a male, will be likely to butt against them and start the nails, which cannot, however, be done when they are backed by the cross rails.

The next operation is to provide the feeding-board mentioned before, and to cut circular holes in it, opposite the openings in the stalls, to receive the feeding-pails. This board (*e*) rests on brackets, either of iron or wood, screwed against the posts at the necessary distances. When the stalls are away from the front wall a rail should be nailed along the edge to prevent anything like lumps of rock salt or whole roots from falling off. All that now remains to be done is to fix the hay-rack. This (*f*) I need not describe beyond giving the dimensions. I may, however, state that I use iron rods, ⅜in. thick, placed 1½in. apart, in preference to the ordinary wooden bars. The width of the rack should be 18in., the length, of course, depending on the number of stalls. When these are placed against the wall the lower part of the rack will be fixed against it, but if there is a passage between to allow of feeding in front, according to my own plan, some support is necessary.

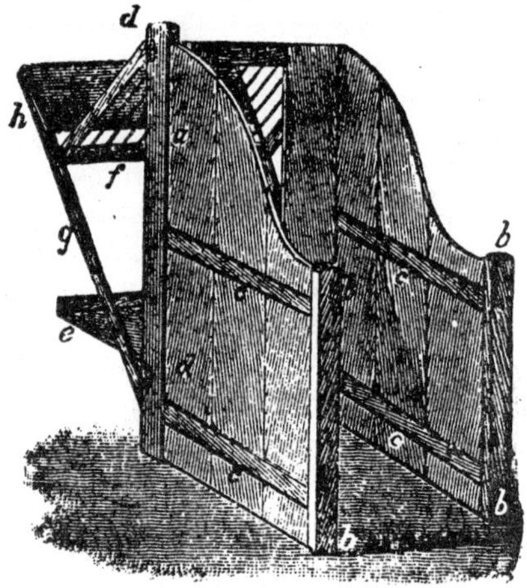

Fig. 4
BACK VIEW OF GOAT-STALL

This may be obtained by nailing a length of wood obliquely, as shown in Fig. 4 at *g*, and fastening the rack against it. The part above, where it joins the rack (*h*), is boarded lengthwise with a couple of thin boards, which extend the whole length, and serve to form, with the rack, a kind of trough to hold the hay. My reason for placing the rack outside the stalls instead of inside is to avoid the waste of hay that takes place when the rack is situated over the animals' heads, owing to portions falling at their feet and being trodden on.

When placed outside, whatever may fall whilst pulled out drops on to the board and in the pails, and is subsequently eaten, being then unsoiled.

Fastenings for Stalls

For fastening up the goats a staple may be driven about the centre of the stall, and 12in. from the ground, but what is better is an iron rod 2ft. long and not quite ½in. in diameter, bent at right angles at each end, as shown in the illustration (Fig. 5); the ends, being pointed or wedge-shaped, are driven into the wall or partition, the lower to within an inch, and the upper within 3in. of the angle of the iron, thus giving the rod an oblique position. Into this rod, instead of the staple, the spring-hook connected with the goat's collar can be slipped, it having this advantage over the latter, viz., that, instead of being fixed, and therefore curtailing the distance when the animal stands up to feed, the hook, as soon as the goat tightens its chain, slides up the inclined rod and gives it the same length as when lying down. Where any difficulty is experienced in obtaining a bar of iron, an ordinary iron curtain-rod answers the purpose very well, the trouble of bending it being avoided by driving a long screw through the eye at the top and a short one through the bottom eye, and fixing it to the wall in that way.

There is yet another plan, closely resembling the above, but considerably stronger, and one which I should recommend for large and powerful goats; it consists of an ordinary screw-bolt (Fig. 6) dropped into a couple of screw eyes and secured on the other side by a nut, a ring having been previously slipped over the bolt to slide up and down; this I prefer to the bent rod, as it is stronger and more easily removed when required. These screws are best inserted in one of the *front* boards of the stall, within an inch of the opening admitting the goat's head, the bottom screw being placed about 12in. from the ground. It is necessary for the goat to be fastened as near to the centre as possible, to prevent her turning round. For the same reason the chain and spring hook attached to the collar should not exceed 10in. in length.

Those who wish to carry out the work at as cheap a rate

as possible should purchase some empty American bacon-boxes sold second-hand by wholesale provision dealers. The boards of which these are made are just the right height and thickness, and, being "matched," may, if the boxes are taken to pieces carefully, be put up for the partitions and fronts of the stalls,

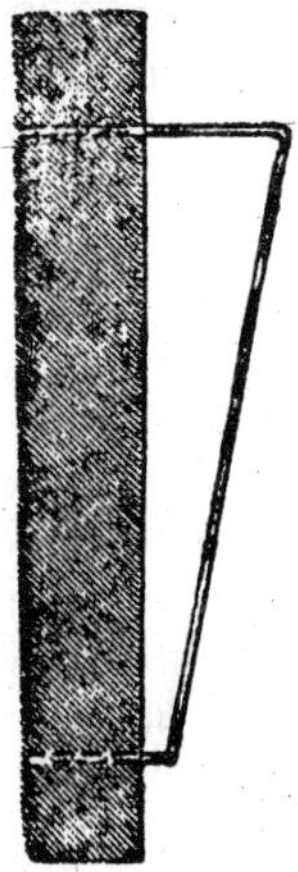

Fig. 5
SIMPLE FASTENING-IRON FOR USE IN GOAT-STALLS

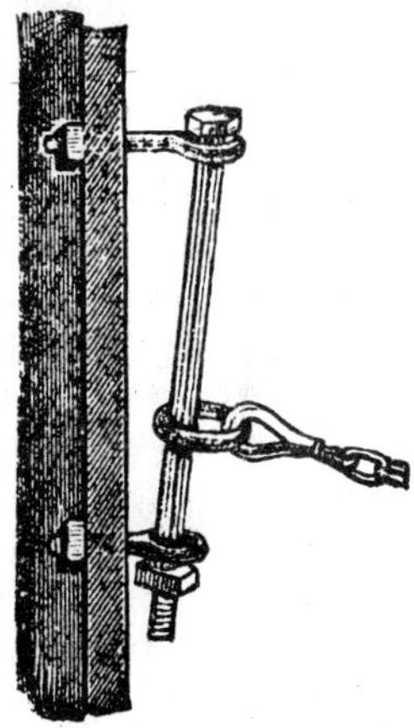

Fig. 6
SCREW-BOLT FASTENING

and be made to look quite as well as new wood, especially if coated with varnish.

SPARRED FLOORS

In some goat-houses it is the practice to use sparred floors, on Mr. Bryan Hook's plan, by which the goats are raised from the ground some 4in. or so. Litter is here dispensed with, but a little peat-moss, finely broken up, should be placed beneath the floors to absorb the moisture. The objection to these floors is the extra labour entailed, as they require to be placed in the open air every day to purify whilst the goats are at pasture, and once a week they should be scrubbed.

PLAN OF GOAT-HOUSE

In order to assist any reader who is desirous of either building a goat-house or adapting some already-erected structure for the purpose, I give a plan showing the most advantageous and commodious arrangement for a stable, based on the system I have myself adopted. The interior of the building is 19ft. long by 14ft. wide, and it is designed to accommodate twelve goats, besides a limited number of kids. *a,a* show the stalls 2ft. wide, and a^x, a^x, those of larger size. Each partition is 2ft. deep, it being merely necessary to extend them two-thirds of the length of the goats' bodies to prevent their fighting, thus giving them plenty of room to lie down in any position, but preventing their turning round; *b,b* show the feeding-boards in front with circular holes for the pails; *c,c* the loose boxes for kids, available also for goats that are kidding, the feeding-boards and hay-racks being here placed corner-wise. Between the two sets of loose boxes is the milking bench, raised 2ft. from the ground, and with the sliding "guillotine" arrangement for fastening the necks of the goats when being milked. This is described and illustrated in the chapter on Milking; but I should here observe that when I first adopted the apparatus many years ago I kept only two goats, and it was then part of the arrangement of the manger to which the animals were fastened. I have since found it better to make a separate compartment for this, and instead of using it with the goats standing on the ground, to fix it on a bench, and to make each goat mount the bench alternately to be milked; this is easily done, and it saves much inconvenience and labour to the operator. The space c^x is for the milker to sit, and on his right hand, and abutting against the adjoining loose box, one or two shelves should be placed to receive the vessels used in the operation, they being then within easy reach; *e, e,* are the windows; *f,* the door, this being, like the passage it leads into, 2½ft. wide, sufficient for wheeling a barrow when cleaning out the stable. Enough space is provided between these and the walls for the barrow to pass when the goats are in. The feeding gangway is made 2ft. wide, to enable the attendant to walk between the stalls at the goats' heads when feeding and watering them. I should here remark that under this arrangement it is necessary for the pieces of quartering or thick rails used for supporting the hay-racks to be fixed upright instead of obliquely, as shown at *g*, Figs. 3 and 4, in order to allow more room for a person to

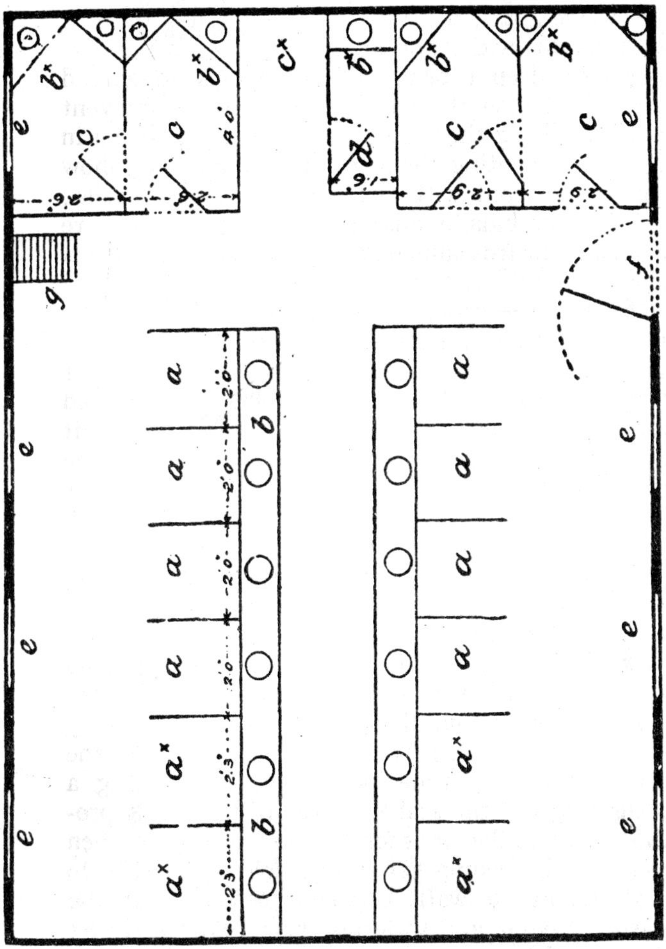

FIG. 7—PLAN OF GOAT-HOUSE

a, ax, Goat-stalls; b, bx, Feeding-boards; c, c, Loose Boxes for Kids, etc.; cx, Milking Compartment; e, e, Windows; f, Door; g, Ladder Leading to Loft. Scale ¼ in. to a foot.

pass down the feeding gangway; *g*, in the plan, shows the ladder leading to the loft when the structure is on a more pretentious scale. A loft is always an advantage in a stable, for besides the convenience it affords for storing hay and roots, cutting chaff, etc., the opening or trap-door is valuable as a means of ventilation without a draught. The entrance *f* should be provided with double doors, so that the top half may be left open whilst the bottom one is shut.

In cases where a goat-house of this description is actually erected—on a goat-farm, for instance—another door should be provided either where the step-ladder is now shown at the other end of the passage, or else between the loose box and the milking-bench, c^x. One of these doors should then lead to an enclosed yard with a hard dry bottom, whilst the other opens on to the pasture-ground.

Compartment for Kids

Besides the stalls for the goats, it is of the greatest importance that some provision should be made for confining the kids when separated temporarily from their dams. Such should consist of little loose boxes about 2ft. 6in. wide by 4ft. or 5ft. long and about the same height, or sufficient to prevent the little animals from jumping over. They are best made of narrow boards.

In days past the best arranged and most costly goat-houses were those of the late Mr. B. Ravenscroft and the still more perfect erection, the joint property of Lady Gertrude Crawford and Lady Arthur Cecil, at Lymington. At the present time the goat-houses with the latest improvements that I have visited are those of Miss Pelly at Theydon Place, Essex, Mrs. Arthur Abbey at Didgemere Hall, Roydon, Essex, Mrs. Browell at Feltham Vicarage, Middlesex, Mrs. Morcom, the Clock House, Bromsgrove, and the most recent of all that of Miss J. Port at Twisly, Catsfield, near Battle, Sussex.

The photograph of the interior of the goat house at Lymington shows on one side the open stalls and on the other the loose boxes, but the former can be also enclosed when desired by light gates, which, when not in use, are placed against the wall of the building. The collars and chains are shown hanging against the partitions of the open stalls. A new arrangement for fastening and unfastening the collars has been designed by the owners, which is extremely simple and most effective. The goats are all fed from the front, by means of a movable frame holding the feeding and drinking pans, the covers for which are seen hanging above. Their position can be faintly detected in the illustration by the fine white lines. At the far end on the left is the milking lobby fitted with milking-bench and the various dairy utensils, some of which are shown. There is here a gas cooking apparatus for warming the kids' food. Further to the right, by the gangway near the wall, is the door leading to the store-room for hay, corn, straw, etc. The goat-house is also fitted with an arrangement for measuring the height of goats, and a machine to obtain their weights, whilst cards and milking scales are provided in the lobby for recording the yields of each goat.

Litter for Goat-house

Several kinds of litter are available for goats, but this is required more as a means of absorbing the excrement than with the object of conducing to the animals' comfort. Indeed, goats appear to prefer being without it, for instead of seeking warm beds to lie upon, like dogs, pigs, and many other animals, they give preference to hard but dry ground ; and when straw, bracken, or similar litter is provided, they generally scrape it away with their fore feet to the rear. For the purpose of collecting and utilising the manure, however, it is necessary that some kind of bedding should be provided, and the more absorbent it is the better. Cheapness is the great point here, generally speaking, and whatever is most available is usually preferred. When straw is at a low price it answers well enough in winter, but should not be employed in summer. In hot weather the goats seem to have a decided objection to being enveloped in straw bedding, and the rapid evaporation of ammonia from the droppings makes the air of the stable unpleasant.

Where a common is close at hand, and dried brake-fern, or bracken, can be obtained and stored for winter use, it answers admirably; and being obtainable for merely the trouble of collection, has no drawback on the score of expense. Sawdust, when easily available and cheap, is cleanly, cool, and absorbent; but it does not make good manure. In summer, dry soil is very good, acting both as an absorbent and deodoriser, and is easily got rid of by digging into the garden on removal from the goat-house. (The objection to this is, however, that it forms into mud when wet and makes the udder dirty and gritty.) Straw or bracken, on the contrary, requires to be collected in a heap to ferment, becoming a nuisance, or at least being regarded as such, in a closely-inhabited neighbourhood.

The best substance of all for litter is dried peat; it has the advantage of being absorbent to a considerable degree, at the same time that it acts to a certain extent as a deodoriser. It is easily stored for use, is cleanly, and has the merit of being particularly valuable for garden purposes. Peat is easily broken up into small pieces, whilst these get further pulverised under the animals' feet, at last presenting the appearance of a kind of coarse, chaffy dust. This may be turned and re-turned several times before it is thoroughly saturated and fit for manure, and even then may be re-dried in the sun and used once again; thus it goes a long way.

CHAPTER XII

PASTURING

THERE are two methods of treating a goat; one is by a system of pasturage combined with stall-feeding, and the other by what is known as "soiling"—that is, keeping the animal constantly housed, and supplying it with all its rations in the stable. The one to be adopted by the goat-keeper must of course depend upon whether he has pasturage at command or not.

By the word "pasturage" let it not be supposed that I mean a small grass-plot, even if it be the size of a tennis-lawn. It is not that the dimensions in such a case would be necessarily insufficient to keep a goat during the summer months, but that after a short time the animal would probably refuse to crop it. As a matter of fact—and this I have proved to my perfect satisfaction—unless a person has a good run of pasturage, and that of the right sort, a goat will thrive better and live longer under the stall-feeding system. Knowing, as everyone does, that this animal is by nature active and restless, constantly roaming about from place to place in search of fresh food, such an assertion as the one I have just made will probably cause some surprise. I have not, however, come to this opinion without a thorough trial; and I have the experience of others as well as my own to justify it. It is further borne out by what is done in the Mont d'Or, one of the chief goat-keeping districts of France. There a great number of these animals are maintained entirely in stables without so much as a day's grazing in the whole of their lives, yet they thrive well, give a quantity of milk (which is utilised for cheese-making), and live for many years. I have had opportunities of comparing goats kept in the country by persons who have a garden with an acre or so of grass with other goats in London, where they are nearly always in a stable with merely the run of a mews or a back-yard, and the latter always seem to be less subject to disease and to live longer. The only drawback where there is no grass or garden is the extra cost of feeding. No one, however, need hesitate about keeping a goat if he has a dry shed or stable in which to place it, and a good-sized garden to supply some green food. This latter is an important adjunct, for a goat should have *some* succulent diet if a large yield is to be obtained, although it does not require to be pastured. When all has to be bought, however, it comes expensive.

There is a great deal of misconception in regard to pasturage for a goat, as I have already hinted. It is commonly said that a goat will live and thrive where most other such animals would starve; this refers no doubt to goats that live a semi-wild life on steep escarpments and elevated plateaus of mountain ranges where herbage is short and sweet, though scanty; but the converse also holds good, that on a rich pasture where cows would yield abundance of milk, and sheep fatten, a goat would be very likely to sicken and die.

A goat loves grazing, but it requires constant change of pasturage, and it soon becomes ill if repeatedly put to feed on the same plot of grass; hence it is that a lawn or an orchard which is too small to allow of such changes being made does not suit these animals. They may do well on it for a certain period, varying from three months to a couple of years, according to the extent of the pasture and the nature of the soil and grasses; but if tethered time after time on the same ground, although it may meanwhile have been washed by rains and refreshed by a new growth of herbage, the goat will eventually become ill. These animals do not care for pasture that has been freshly manured, or that has had the droppings over it, to any extent, of pigs and poultry; they like coarse grass if not of rank growth, but the kind they prefer, and which suits them best, is the short sweet pasturage on downs and dry commons. On the latter they have the great advantage of a change from the grass to the sweet and tender shoots, prickly though they be, of the furze or gorse bushes, of which they are extremely fond, and which, moreover, contain a considerable amount of nourishment. Indeed, although a goat, as I have said, enjoys grazing, it prefers browsing, the height of its felicity being reached when it can obtain both at will, for it is the constant change from one kind of grass to another, and from cropping herbage to nipping off leaves and shoots, that these animals delight in.

It is this peculiar fancy for biting off tender buds and barking trees, and its innate love of destruction, that render the goat such an enemy to the gardener and the farmer, and make it such a disagreeable, not to say expensive, matter to the owner when one of these mischievous creatures makes its way by accident into his own or his neighbour's garden or fields. The only preventive in such cases is the use of a tethering-chain and pin, as I shall presently describe.

An acre of grass is the least that can with advantage be allowed for two or three goats in order to give them a frequent change of "bite," so that they do not go over the same spot many times in the course of a season. But even then the pasturage should be varied with leaves and other food for a day or more at times. This change should be given on wet days, for goats do no good tethered out in the rain ; they hate wet, if it be even a slight shower, but in a heavy storm of wind and rain they go nearly frantic, running round their tether in a furious manner, bleating loudly all the time, and generally managing in the end to loosen their tethering-pin sufficiently to be able to get free, when they rapidly seek shelter. A goat does not come to the harm on wet grass that a sheep will, being less subject to the complaint known as "hoven" (caused by an excess of green-stuff in a saturated condition) than the latter ; at the same time it is not good for them, neither is the grass that has been pushed into rapid and sappy growth by a series of soaking days in a warm atmosphere. It is far better to mow down any over-luxuriant verdure of this kind on a change to fine weather, and to feed this in small quantities, with hay and corn, to the animals when in the stable. Where a goat and a cow are put on the same pasture economy is effected, for the cow with its sweeping tongue soon takes off the overgrowth, leaving the goat, which can bite much more closely, to follow it after the rankest has been removed.

It is a great mistake to put a goat for long on pasture after being accustomed to dry food such as hay and corn, more especially in the spring of the year, as the fresh grass is sure to scour it, and sometimes so seriously as to cause death. The change should be made as gradually as possible, starting with, say, a quarter of an hour or so in the morning and afternoon at first, and extending it by half an hour a day until the animal is thoroughly accustomed to the new diet. Pasturing should not be attempted before the middle of March, and should cease altogether in the middle of September. An occasional run over the field for about an hour now and then on fine days, and when the ground is tolerably dry, is all that should be permitted during winter, and then only after a meal of hay or corn. Avoid pasture on which sheep have run.

Soil and Grasses

Another important consideration in regard to pasturage is the soil. Goats seldom do well on a stiff clay and never on a marshy kind of ground, so that, except during the driest days of summer, it is far better not to put them on grass at all if the soil is of this description. Gravel will do very well, but the best soil is chalk. Besides the comparative dryness of the latter, the herbage that thrives upon it, and which at the same time never grows to rank luxuriance, is the kind of which the goat is most fond. Amongst this may be mentioned the *Festuca ovina*, or Sheep's Fescue, a short, fine grass, which grows in a tuft at its roots, and pushes up delicate stems rarely exceeding 12in. in height. This grass abounds on the highlands of Scotland, the mountainous parts of Wales, and on many of the downs in England. Two other kinds of the same genus, viz., the Hard Fescue (*Festuca duriuscula*) and the Red Fescue (*Festuca rubra*), are also favourite grasses of goats. When a goat is able to roam about and choose its pasture, it can correct any redundance of one kind of herbage, which alone might have a prejudicial effect, by a change to another sort having an opposite tendency. This it is unable to do when tethered.

Tethering

This is performed by means of a chain, one end of which is fastened to a "tethering-pin," and the other slipped into a spring-hook attached to the collar or head-stall of the goat. I must here remark that these animals being very powerful, whatever is used to secure them, either in the stable or out of doors, should be of the strongest kind. A rope is not good for this purpose, for although it may be strong enough for a time, it "kinks" after being dragged through wet grass, shortening the length of the tether thereby, and also getting entwined round the body or legs of the animal. It is much better in all cases to use a light but well-made iron chain furnished with a couple of swivels, one close to the tethering-pin and the other at the end where the chain connects with the collar. These

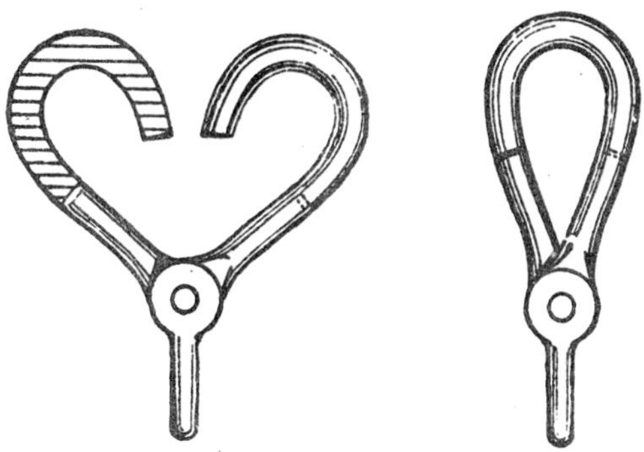

Fig. 8
SPRINGLESS HOOK OR DOUBLE SHACKLE FOR TETHERING—AN IMPROVEMENT ON THE SPRING HOOK

prevent the chain from getting twisted into knots. Suitable chains of any desired length may be procured at all large ironmongers' at a cost of about sixpence per yard. A dog-chain is hardly long enough for this purpose, as the length of the tether should be quite 3yds. Tethering-pins may be of various shapes and sizes. That shown in Fig. 9 is a square rod of iron, 15in. long, tapering to a point, the head of which is furnished with an "*s*-hook," which will turn in any direction without entangling the chain. This can be made by any blacksmith, and costs about two shillings. The chain and *s*-hook are best connected by means of a springless split hook which at this part is better than the spring hook, as the latter is apt to get clogged with dirt, when it fails to act, or the spring becomes displaced, in which case the goat gets loose. The springless hook is so constructed that it is impossible for it to become accidentally detached or to get out of order. It may be used in place of the spring hook at the collar end of the tether if desired for perfect safety, but it takes rather longer to disconnect than the latter.

When two or three tethering-pins are required, they may be made at much less expense by procuring a similar rod to that

Fig. 9
TAPERED TETHERING-PIN

Fig. 10
BENT TETHERING PIN

described for fastening the goat in the stable. One 4½ft. long, which may be bought for about 1s. 6d., will make three pins. The rod is easily broken into three by first filing all round the part where the fracture is required; then take each piece, and after heating to redness, bend one end in the form of a loop, as shown at Fig. 10, and hammer the other extremity to a point to facilitate its insertion in the earth. The length of the pin from the loop to the point should be quite 18in. This kind of tethering-pin has two advantages over the previous one. In the first place, if the loop be made the exact shape of that shown in Fig. 10, no swivel will be needed, as the pin turns round in the hole it has formed in the ground to whatever direction the goat moves, and, in the second, it is easily forced into the earth by the pressure of the arm, whereas the square rod requires to be driven in with a mallet, which is not always at hand. The square pin is, of course, the stronger, but it has this objection that if struck too hard with an iron hammer the head is very apt to break off. Care should be taken in making it that the s-hook works freely round the neck of the pin, which should be driven with the head level with the ground.

A Cheap Form of Tether

A very cheap and effective tether (Fig. 11) may be contrived as follows: The peg (P) is made of ½in. round iron, with an eye turned and welded at the upper end and the lower end drawn out to a dull point, a swivel (S) being attached to the peg by means of a welded ring (W). A piece of good rope (R) or a light chain is fixed to this, and a toggle (T) attached to the

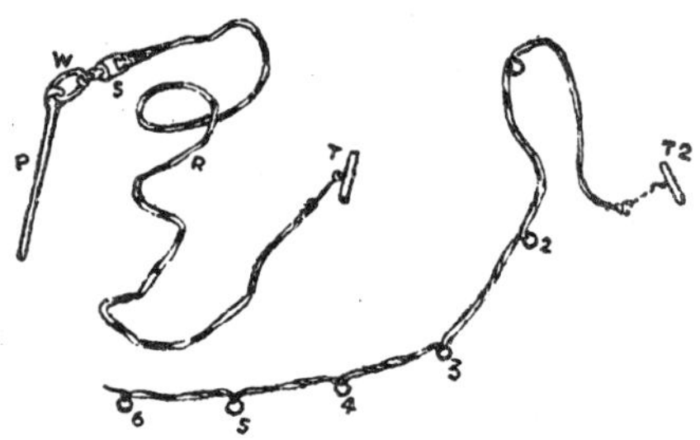

Fig. 11
Goat Tether and Extra Rope

other end of the rope, when the tether is complete. The length of the rope should be at least 15ft., and that of the peg about 12in., but that depends on the nature of the soil. For heavy soil, and during the summer months, a shorter peg is better. The peg should always be driven with a wooden mallet, and must of course be driven into the ground down to the ring, so that there can be no possibility of the rope twisting round the peg. Now attach a ring (suited to the toggle) to the goat's collar, and you are in a position to fix the goat to the tether and to release her at night with the least possible trouble.

If a second toggle is suitably fastened in the goat's stall or stable, she may easily be fixed there. Such a tether may be made for 1s., with a special swivel. The frequent drawing and driving of the peg may be avoided to a great extent in the following way: Procure another piece of rope, fix another toggle to one end of this, and tie, say, six rings into the rope, as in the figure. Pass the toggle T through ring 1, and toggle T2 through the ring on the collar. Now, instead of moving the peg, all you have to do is to take toggle T out of ring 1, and pass through ring 2, and so on, a matter any child can attend to. The distance between the rings should be less and less as the rope increases in length (thus, the distance between 6 and 5 will be less than that between 2 and 1), otherwise the goat will get a larger bite as the rope increases in length, the circle being greater. This novel form of tether used to be procurable years ago ready made at the cost of 1s. extra toggles being 2d. each. The rope and ring extension nobody should have any difficulty in making up for himself.

Combined Tether and Shelter

The chief objection to tethering a goat is that the animal is exposed to sudden storms of wind and rain, and it may be drenched before it can be freed from its tether and brought under cover. Anyone who has witnessed the frantic efforts of a goat to get loose under such circumstances—for these animals, as I have said, have a great objection to rain—will appreciate this fact. I have tried various devices for overcoming the difficulty, such as fastening the end of the tether to a kind of dog-kennel structure on wheels, pinning this down to the ground with an iron rod so that it cannot be dragged about by the goat. In nearly all these arrangements, however, the goat manages to get its tethering-chain wound up or caught in some way, and its comparative freedom becomes then curtailed or absolutely annulled. The best plan I have found to effect the object desired is to use 50yd. lengths, or longer, of thick galvanised wire—if expense be no object, preferably that made of a number of strands of thin wire twisted together, as used for clothes-lines. This wire is pinned securely down at each extremity, care being taken that the pins do not project above ground and thus catch and arrest the tethering-chain. The pins in this case should be of the pattern shown in the illustration of the toggle tether, having an eye in which a ring or springless hook is fitted, and the length of the pin should be quite 18in. The tethering-chain used here may be short, say, the length of an ordinary dog-chain, as the springless hook which catches into and runs along the wire as the goat moves gives the whole length of that wire at which to pasture from. At one end of this wire a movable shelter may be placed, at such a distance that the goat can only just enter it, but is unable to get round or upon it. Such shelters may be rough structures or ornamental according to their position in respect to the dwelling-house and to the taste of the owner. When these are used a goat need not during the summer months be taken to the goat-house at all, but may be hand-fed if required and milked out of doors.

Substitute for a Tether

Some people, instead of tethering a goat, let it roam about at will, having previously fastened a "puzzle" round the animal's neck, which prevents it from getting through fences and palings. The puzzle, a sketch of which is given (Fig. 12), is made in the following manner. Procure three pieces of strong wood, or, what is better, if obtainable, three sticks of bamboo, which combines strength, pliability, and lightness; these should be from 2ft. to 2ft. 6in. long according to the size of the goat, and must be fastened together in the form of a

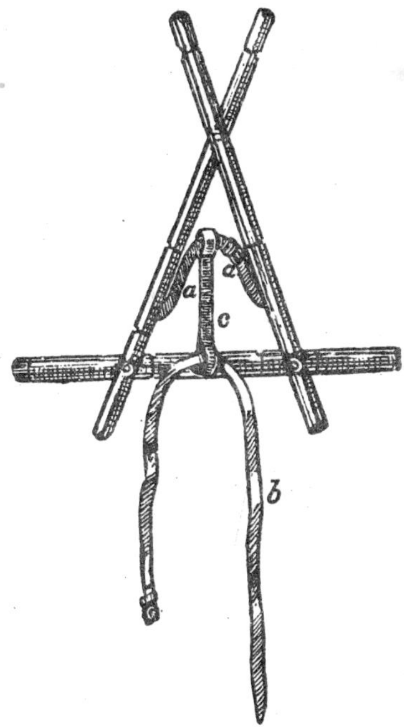

Fig. 12
The Puzzle—a Substitute for a Tether

triangle, leaving the ends projecting. The horizontal piece at the bottom, which is somewhat thicker and heavier than the other two, should be secured to the latter by means of iron pins or rivets instead of being nailed, as this allows the sticks to work freely one against the other. The cross-pieces at the top should be joined by a "thumb" or "fly" nut, in order to be easily parted when the apparatus is put on the animal. When applied to goats having no horns, the fly nut is unnecessary, as, if the pieces are fixed, the whole thing can be put over the creature's head. To the inside of the sticks a couple of straps

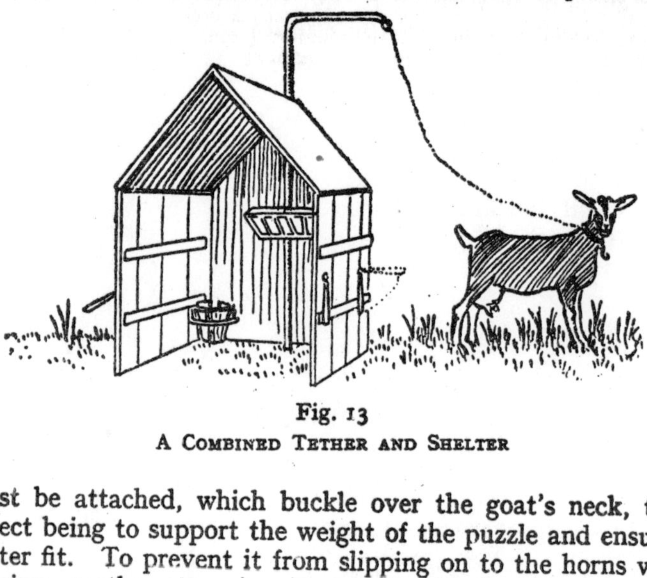

Fig. 13

A Combined Tether and Shelter

must be attached, which buckle over the goat's neck, their object being to support the weight of the puzzle and ensure a better fit. To prevent it from slipping on to the horns when grazing, another strap is necessary, which buckles round the body, and is joined to the former by a shorter piece of leather having a loop at each end through which both pass. The leather handle of an ordinary double rug strap answers very well for this purpose. The arrangement will be better understood by a glance at Fig. 12; *a a* represent the two straps which go over the neck, *b* that which fits round the body, and *c* the short piece joining the two together.

Mrs. Morcom, of The Clock House, Bromsgrove, has invented a novel combination of tether and shelter. The tether is attached to a long pin, which revolves in the hole provided for its reception as the goat walks round the shelter, and never gets wound up. This will be understood at a glance by the accompanying illustration where the pin is shown driven into the ground through the shelter which contains a hay rack and water pail.[1]

[1] Under the title "A Protection against Goats," etc., the following interesting note appeared in *Notes and Queries*, 9th Oct., 1880, signed Geo. A. Muller, St. Martin Lautosque, Alpes Maritimes: "Nearly everywhere in the Maritime Alps, in a good many parts of Switzerland, in Southern Germany and Tyrol, people who want to protect their land from the inroads of goats, sheep, and cattle in general put up a stick surmounted by a small bundle of straw, or simply tie a small bundle of straw to a branch, and thereby find their ground as safely guarded as if they had placed a policeman there. Is this custom to be found anywhere in England? I fancy I have noticed it in North Wales. What is its origin?"

CHAPTER XIII

STALL-FEEDING

In feeding any kind of milch stock, there is this question to be considered: What description of food will produce the largest return in milk? From some carefully conducted experiments made some years ago in Germany to ascertain the influence of different foods on the quality and quantity of milk, the conclusions arrived at were that the quality of milk, in so far as the relative proportions of its constituents are concerned, is but slightly affected by any particular diet, but depends on the breed and peculiarities of individual animals; the quantity, however, may be affected by the nature and character of the nourishment taken.

Without going into scientific details, I may just mention that every species of food, as all are aware who have any knowledge of chemistry, contains two important elements, the nitrogenous or flesh-forming, and the non-nitrogenous, or heat- and fat-giving, otherwise called carbohydrates. Now, it is to ascertain which of these two elements has the greater influence in the production of milk that most of the experiments have been directed. In the case just referred to, the largest return of milk was furnished by a diet rich in nitrogen, the food supplied being at one time highly nitrogenous, and afterwards containing but a small proportion of the flesh-forming element.

Other investigations tend to show, however, that the nature of the food consumed does not exert so very sensible an influence on either the quantity or the chemical composition of milk if an equal amount of nourishment is obtained from the different kinds of food. During the last half-century chemistry has made such rapid strides, especially with regard to analysis, that at the present day there is scarcely an article of diet in its simple form the chemical composition of which has not been ascertained. The following are the cattle foods which contain the largest amount of nitrogen: First and foremost stand the various cakes made from linseed, rape- and cotton-seed; then in their order follow lentils, beans, peas, clover-hay, bran, and oats, and among the fodder plants grass (timothy and meadow foxtail), tares, lucerne, red and white clover.

I mention these particulars in order that those who desire to experiment upon their goats, with a view to ascertaining the relative influence of the different foods upon the produce of the milk, may be enabled to do so. For my own part, I am quite certain that however much the chemical constituents of the diet may affect the supply of milk, the *quantity* of food assimilated by the animal influences it much more. It is generally agreed by dairy-farmers that grass and roots, but grass particularly, make more milk than dry food. This I have also found to be the case in feeding goats, and I attribute the increased yield to the succulent nature of the diet. Experiments made some years ago in feeding milch cattle showed that dry food operates in the direction of an augmented consistence of the milk and an increase in the live weight of cattle fed upon it, whilst green food, on the other hand, chiefly influences the quantity yielded. The foods used in the experiments were green clover in the one case, and clover-hay in the other, combined in both instances with oat straw.

As regards goats, however, the great secret in obtaining a large yield consists in a constant change of diet. I believe there exists no animal, if we except the pig, that eats so great a variety of food as the goat. Yet, on the other hand, there is none which sooner tires of one thing, or is more fastidious in regard to the cleanliness and quality of each article.

Green Food

With regard to the variety of plants a goat eats, some experiments were once made in Sweden and in France which bear on this point, and I therefore give them for what they may be worth.

Experiments in Sweden

Of 575	Plants the	Goat	eats 449,	Refuses	126
„ 528	„ „	Sheep	„ 387,	„	141
„ 494	„ „	Bull	„ 276,	„	218
„ 474	„ „	Horse	„ 262,	„	212
„ 743	„ „	Pig	„ 572,	„	171

Experiments in France

	Goat	Sheep	Bull	Horse	Pig
Can eat	547	408	311	268	86
Is very fond of	28	81	121	113	36
Sometimes eats	32	33	70	39	23
Takes in all	607	522	502	420	145
Refuses	83	133	183	235	169
Total plants examined	690	655	685	655	314

The results are, apparently, not alike in both cases, owing, probably, to different herbs abounding in the two countries. As in France the experiments appear to have been carried out in the greater detail, we may take the latter to be the more correct; there we see the goat consumes no fewer than eighty-five more plants than the sheep and four times as many as the pig.

I should be careful to observe, however, that there is a vast difference between what goats as a race will eat, and what an individual specimen is ready to consume. Any person keeping half-a-dozen of these animals together will find each has its likes and dislikes in regard to its food, quite as much as any member of his family, one eating with avidity what another will reject. In fact, that essentially goaty word "capricious" applies as much to its appetite as to other characteristics, and therefore, it is impossible to lay down any hard and fast rule as to what a goat may be fed on, and expect it to apply to all goats alike. There are some things which I may mention as generally supplied in the way of green food, and which are in most cases appreciated, even though a small portion only may be consumed at one time.

Roots.—These in their order of preference are: mangolds,

swedes, potatoes, carrots, Jerusalem artichokes, parsnips and turnips, the last-named being liked perhaps least. They must all be scrupulously clean; though in regard to mangolds and swedes it is less important if they are cut in half, and each hemispherical portion placed, the cut part upwards, at the bottom of the goat's pail. It can then scrape or bite off portions with its teeth without the root being displaced by the force and action of the jaws. The operation will be continued all around the inside until only the rind is left, like an empty shell. I have tried giving roots pulped in a machine, and also the same cut into squares, but the other plan answers best by far, being less wasteful, and at the same time it appears to give more enjoyment to the animals. Mangolds should be given chiefly after Christmas, for when first pulled they are less nutritious and very apt to scour. It is best in all cases to begin this kind of food gradually. Small potatoes called "chats" may be supplied whole, but these, as well as larger ones, are preferred chopped. They require to be carefully washed free of all grit first, otherwise they will be refused. Carrots and parsnips are nice wholesome roots, and, containing as they do a large quantity of saccharine matter, are fattening; the peelings of these and potatoes are generally eaten as readily as the roots themselves.

General Garden Produce.—Amongst other garden produce may be mentioned the haulm of peas and beans, lettuces run to seed, cabbage-stumps, the early fallings of pears and apples when too small to be utilised in cookery, the thinnings of raspberry-canes, ivy cuttings, the prunings of all fruit-trees, the leaves even of dry autumnal fallings, and the trimmings of hedges and vines; lastly, acorns and horse-chestnuts. No goat ever refuses the produce of trees, such as fruit or branches: they have an especial preference for the bark and peel, and even eat the wood itself when of the first year's growth. Oak, elm, ash, lime, willow, birch, hazel, vine, and all fruit trees are consumed with avidity. It is said that the peelings of osier and the bark of the willow are a cure for cachexy in the goat. The falling leaves of the various trees above-mentioned are economic food, and much appreciated, as any goat-keeper will soon discover if he watches these animals devour each leaf as it falls in an orchard

Cabbages and lettuces will be consumed almost to the base of the stalks if they are hung up at an elevation so that the animal may constantly nibble without soiling any portion with its feet, otherwise only a small proportion will be eaten. This applies generally to all the food of a goat, for the moment it is trodden underfoot it ceases to be eaten. It is therefore a great advantage to place branches or fibrous-rooted vegetables in the position I have described, for otherwise a large quantity will be wasted.

Acorns are in some years very plentiful, and as they can then generally be bought at about 1s. per bushel in country places it is advisable in such circumstances to lay in a stock. Besides being a cheap kind of food, they are useful in a medicinal way, as, having a constipating tendency, they are valuable as a change when the bowels are relaxed through the animal having partaken of food with an opposite effect. This binding effect is less marked when acorns have sprouted. In that condition, saccharine matter being developed, they are rather fattening and highly relished. All that is necessary to produce this germination is to pile them in a heap a few at a time in a warm damp place. They should be washed clean first, for if this be done afterwards the sweet, tender shoots which the goat so much enjoys are knocked off in the process and lost.

Weeds.—There are many weeds that goats will eat readily, and which are generally consigned to the rubbish heap. As instances of some of the commonest I may mention sow-thistles, docks, plantains, early spring nettles, and dandelions. They are also very fond of meadow-sweet, though cattle will not touch it. All leguminous plants, as vetches or tares, lucerne, sainfoin, clover, and such like, are eaten by goats, but not voraciously, these being more succulent herbage. Though good for the production of milk, they do not, however, appear to be appreciated by these animals, who give preference to hard, bitter, and woody kinds of plants. Amongst these last is tansy, which is apt to impart a flavour to the milk.

Trees and Plants to be Avoided.—There are many plants and shrubs that are poisonous to one kind of animal and devoured with impunity by others, goats being generally amongst the latter. Thus, the long-leaved water-hemlock will kill a cow, whereas the goat browses on it greedily. There are a few things, however, which should be avoided, as they are generally poisonous. I say, generally, because, strange to relate, even the same shrub does not seem to have at all times an equally prejudicial effect; thus yew will cause death to sheep, goats and other ruminants at one part of the season, but at another may have scarcely any ill consequences whatever. This is one of the trees, however, which goats should be kept carefully away from. Another is privet; the leaves have not, I believe, any ill consequences, but the berries are apt to prove fatal. Monkshood and rhododendrons have also proved very poisonous to this class of stock. I have read that pods of laburnum, laurel leaves, and bladder senna are fatal to goats, but I have had no such experience of these. Some people imagine that ivy is bad for goats, but it is not so, and these animals are very fond of it. Some breeders speak strongly against ragwort, a weed with brilliant yellow flowers. Some goats eat it without harmful effects, but others either will not touch it or consume it with bad results. It causes severe scouring in young cattle.

Dry Food

Hay.—Hay is best supplied in its entire state; but it may be cut up into chaff and mixed with the corn. Anyone possessing a chaff-cutter may find it worth while to chop up hay and oat straw together when a number of goats have to be fed, but bought chaff is generally inferior stuff, and not so good for milking-goats as hay. This bulky food serves the purpose of filling the stomach, which requires a certain amount of distension to enable it to perform its functions properly. To effect this with corn alone would, in the first place, be expensive; and, secondly, so large a quantity of concentrated food would be injurious. In fact, hay or chaff given in conjunction with corn may be regarded in the same light in the diet of a goat as bread and vegetables combined with meat in that of man. There are several kinds of hay, but that preferred by a goat is composed mainly of grass with a few herbage plants, and lucerne hay, made with that plant alone. Sainfoin hay is also good and very nourishing, but more wasteful, as the leaves drop off and fall on the ground.

Oats.—Oats may be regarded as the staple corn for goats, being more frequently given than any other. They are, moreover, the cheapest of all grain when the quality is good—cheap oats are useless, being nearly all husk. Good oats are clean, hard, heavy, sweet, plump, and rattle like shot. The hard pressure of the nail on an oat should leave little or no mark, whilst the kernel when pressed between the teeth should chip rather than tear. Short plump oats are preferable to large long grains. The best weigh from 40lb. to 42lb. the bushel, whereas inferior kinds weigh only 30lb. or 32lb.

Maize, or Indian Corn.—This grain contains a considerable amount of nutritive properties, the greater proportion of which go to the formation of fat; but it should be given kibbled, being rather indigestible. In this way a little is good for a change, but given in large quantities and with no other food, it may cause inflammation of the stomach, which often ends fatally. There are two kinds, the large and the small, the latter being the more suitable, though rather dearer. The large kind should always be given crushed.

Beans and Peas.—Both beans and peas, as I have already stated, contain a large amount of nitrogen, and therefore make valuable food for milk production. Given in the form of meal, beans have the character of improving the quality of the milk, but I do not find that goats take readily to this; besides, it comes rather expensive. Beans should be split; they are most nutritious when about a year old. In Algeria goats are fed largely on beans, and the enormous yield of the Maltese in that country is attributed greatly to this.

Wheat and Barley.—Goats eat wheat and barley eagerly, but these grains are not good for them, as much passes through the alimentary system undigested. Barley, both in the grain and as meal, is too heating, and is apt to cause an outbreak on the skin.

Miscellaneous Food

Under this title may be mentioned brewers' and distillers' grains, and all sloppy food, also oil-cake, bran, middlings, etc. These have a very decided effect upon the supply of milk, the grains and sloppy food influencing the quantity, and the cake, etc., in a slight degree the quality. Many goats will not touch grains ; most of them prefer the brewers' to the distillers', though they may be brought to like both by being tempted with small quantities at a time, and especially when they see other goats eat them. Some goats have a great partiality for sloppy stuff, like bran mashes, coarse oatmeal upon which boiling water has been poured, or middlings mixed with water or skim-milk. In the Mont d'Or, where, as I have already stated, goats are largely kept on the soiling system, this taste is encouraged, greatly to the advantage of the goat-owner. Here the owners even go so far as to mix the meal with pot-liquor. There is no doubt, as I have frequently proved, that where a goat has a liking for a sloppy diet it increases her milk supply materially.

Oil-cake, linseed and locust bean meal, and such like luxuries make a nice change of diet, but they should be given sparingly. Cotton-cake is cheaper than linseed-cake, but less digestible. It has a constipating effect, and may, therefore, be given when slackness of the bowels is observed. Compound cake, I find, goats eat with the greatest avidity.

Water and Salt

Besides dry food and forage, there are two other articles of equal importance which must never be omitted ; these are water and salt. The quantity of water a goat drinks depends upon the amount of milk she is supplying, and the nature of the food she consumes. One that gives 2qts. of milk a day and exists chiefly upon hay and corn may take in summer 2qts. to 3qts. of water in twenty-four hours ; whereas the same animal in a dry state, and living on green and succulent vegetables or grass, will not drink as much in a week. Goats should never be allowed to suffer from thirst, but should have a pail of clean water offered them morning and night, and if rejected, it should be left by their side for a quarter of an hour or so (but not permanently), as although they may refuse to drink before having had their feed of corn, they will often do so when the latter has been consumed.

It is a bad plan to leave water standing in the goat-house all day, as it is liable to become tainted by absorbing the ammonia and other gases from the litter, in which state it will probably be refused. Water, too, that is soiled or greasy, or in which chaff or hay has fallen, will be just sniffed at and turned from in apparent disgust; indeed, I have even known goats go so far as to overturn the receptacle that contains it. Where clean soft water is procurable, it should certainly be given in preference to hard. Water is always much appreciated if given warm and especially if a handful of scalded oatmeal is thrown in.

Next to food and water, salt is most important for maintaining health and condition. It assists digestion and furnishes certain necessary elements to the blood. Goats, like most herbivorous animals, are particularly fond of salt, and will lick it with great relish for a length of time, those in a wild state travelling long distances from their usual haunts in order to obtain it. This article should not be given occasionally, but constantly, a lump of rock-salt being left at all times where access may be had to it whenever desired. The larger the lump the better, as small pieces often get lost, or are eaten whole, in which case more is taken into the system than is really beneficial.

At certain periods of the year salt is more requisite than at others; in spring, for instance, when grass is deficient in saline material, a small quantity of table salt may with advantage be mixed with the provender. A good way of giving salt to goats is to hang up in the stall one of the rollers of compressed salt supplied by Spratt's Patent, Ltd. These revolve on a spindle as they are licked.

Rations

The quantity of food to be given must depend on the capacity of the particular animal, some being larger eaters than others. Goats that are in full milk require to be fed more than those that are going dry. According to a French writer, M. Magne, the goats in the department of Mont d'Or receive during summer four meals a day. The first is given at about six o'clock in the morning, the second at eleven, the third at four, and the last meal at eight in the evening, the goats receiving, to begin with, an armful of cut grass, clover, or tares, then chopped roots or peelings of vegetables mixed with bran or ground-up oil-cake and soaked with water, the third meal being dry food of some sort, and the last meal a feed of hay. In winter there are only three distributions, the quantity of water mixed with the food being increased.

For my own part, I cannot say that I find goats generally relish these mixtures made moist with water, probably because they are not brought up to eat this kind of food. It would be better if they were. My own arrangement for the distribution of the meals is somewhat as follows, allowing for certain changes according to what is most plentiful in the way of green food :

In summer, the first feed, which is given between six and seven o'clock, consists of a good armful of cut grass, goats that are in profit receiving also whilst being milked two large handfuls of oats with one of bran ; water is also supplied at this time. About eleven o'clock, leaves or vegetables from the garden. At three in the afternoon, grains or bran, toppings or meal of some sort, mixed with water. Lastly, at eight in the evening, either cut grass or hay, with oats to those requiring same, as in the morning, water being again offered.

In winter the first meal is given at 8 a.m., and consists of hay ; the second at midday, when either vegetables, roots, or peelings are supplied ; grains, acorns, or cake are given at four ; and at eight o'clock oats and hay.

I always make a point of having the goats fed in the evening not earlier than eight o'clock, and on substantial food, as they have to go so long before the next meal in the morning. This arrangement, it must be understood, applies only when goats are housed all day. When pastured, the corn is given still to those goats that are milking and others that are low in condition ; but the rest have hay only.

In feeding with corn it is well to mix some bran, chopped hay or clover with it when the animal is a voracious feeder, but otherwise it is not necessary, as some goats will object to this addition, and then the corn is best supplied alone. When mangolds or potatoes and other peelings are given chopped up they will sometimes be more acceptable to the animals if dusted over with bran, pollard, or middlings ; they are of course more nourishing under these conditions, and the addition is useful for heavy milkers. Bean-meal and peas in these cases have a very decided effect in increasing the yield of milk, and may be given by way of a change; but this food is too expensive for economic use.

Rules in Feeding

In serving out the rations to the goats there are three important rules to remember: First, that whatever the food, it be presented in a perfectly clean condition; that the pails or feeding-boxes be equally clean, and that if there should be the remains of a previous meal left in them, they be thoroughly emptied before placing in anything fresh.

Secondly, that no more be ever given than can be consumed readily at one time. There is no creature that can be more wasteful in its food than a goat if improperly fed. Some of these animals strongly object to hay-chaff being mixed with the corn, and will turn out half the contents of the pail in their endeavour to pick out the latter. The moment a goat is seen to leave off eating readily, and to begin "nosing" its food in order to select tit-bits, the contents of the pail should be removed. Hay should always be supplied sparingly, otherwise half of it will become litter instead of nourishment.

The third rule consists in feeding at regular hours. This is an important point and should be carefully observed. The appetite of an animal is more even and it enjoys its food better when supplied at stated and regular intervals than when food is given at constantly varying periods.

Cost of Feeding

Having at various times carefully weighed and measured the rations consumed by my goats, I am able to state with some precision the cost. For goats that are pastured daily, with the exception of the few days during which they are housed, this amounts to but little more than the proportional rent of the land. During summer, therefore, as far as grass is concerned, the cost need not be set down at more than 6d. per week. This is, indeed, what is generally charged when goats are pastured out on other people's land. When the goats are stall-fed entirely each will consume on an average

2lb. of hay, costing 1½d. (reckoning the price at £6 per load); about 9lb. of roots at 30s. per ton, value also 1½d.; and a feed of oats and bran measured out in two large handfuls, and weighing, the oats ¼lb. and the bran 2oz., costing together about ¾d. Grains, and many other articles enumerated in the chapter on Feeding, I do not take here into account, as when these are supplied a smaller quantity of roots and hay will be required, and the substances I have estimated for are those which will come to the most money. I am at least on the safe side in the selection made. The sum of these items represents a total of 3¾d. Add a small quantity of oil-cake for goats that are in full yield or want feeding up, or for others whose appetites are such that the ordinary rations are insufficient, and roughly, in order to avoid the charge of having underestimated my expenditure, I will set the cost per day at 4½d. A goat that is pastured in summer and stall-fed in winter should not cost during the twelve months more than £4. The cost here given may be much reduced when the owner makes his own hay and grows his own roots and much of the garden produce, as neither hay nor roots would come to as much as is here set down.

Storing Food for Winter Use

Goats are very fond of all kinds of dried fodder, and there are numerous articles which in summer are plentiful enough, and may be utilised in winter if collected and dried during hot weather and then stored. Indeed, many goats prefer the dried food to the same articles in a fresh condition. This plan is largely adopted on the Continent, and the mode adopted is well described by a writer as follows:

"Keep your garden trimmings, your tree, and bush, and hedge loppings from the rubbish-heap while the leaves are yet unturned, and lay them on sheets or flags in the midday hours to dry thoroughly on the stalks and boughs, during which time they must be well watched, and no intervals of damping allowed. Throw them under a roof at night, and out again when the dew is off; then store them up lightly on poles laid across the roof-beams of a barn or other outbuilding, or suspend them in bunches, like pea-stick faggots, from the beams themselves. In this way vine-clippings, the prunings of fruit-bushes, thorn-hedges, rose-trees, and even herbaceous plants, as chrysanthemums, etc., may become a winter store, and by the succulency they retain permit the introduction of more straw into the chaff of a dear winter."

In regard to this kind of food I may remark that when goats are allowed to wander about in a yard as a means of exercise, it is a very good plan to obtain all the loppings of trees you can get, and to throw them down for the animals to peel off the bark.

Grooming

It is a good plan where goats are kept constantly in the stable with little or no exercise to brush them down every morning with a stiff dandy brush, and, if the hair be very long, now and then to comb it out. Some persons may call this absurd, but when they consider that grooming is to an animal what a bath is to a human being, they must admit that the practice cannot be otherwise than beneficial. The advantages gained are indeed more than might be supposed. In the first place, every domestic animal is liable to the attacks of vermin, whether fleas or lice, goats being by no means exempt; on the contrary, and especially when in a poor condition, they are often very much troubled with the latter pest. By the use of a hard brush, well applied every day, the dirt which collects on the surface of the skin, and which seems to suit such vermin, is removed, and, having no longer a resting-place, they disappear—at least to a great extent—the object of grooming being more as a preventive than a cure. When existing in large numbers other means must be used to exterminate parasites. In the second place, the friction on the skin of the goat by the brush causes a quickened circulation of the blood, thus to some degree taking the place of exercise and so promoting health. Finally, the use of the brush gives the coat a sleek and glossy appearance, and by reason of its extra cleanliness allows of the animal being stroked and caressed—marks of kindness to which it is very sensible—without soiling one's hands.

Paring the Hoofs

When goats are kept constantly in stables it is necessary to examine and pare the hoofs every now and then, otherwise they are apt to grow to abnormal proportions. I have seen these animals, when this matter has been neglected for a long period, go about with the horny part of the hoof three or four inches long and turning up at the ends like a Dutch shoe, causing a peculiar rattling noise as it walked. When goats have a daily run on hard ground, the hoof is of course worn down by the friction induced, but if the animal is always standing on straw or other litter there is nothing to prevent the corneous substance from attaining excessive growth. With some goats this growth seems to be more rapid than with others, though kept under the same conditions, and they sometimes require their hoofs to be pared even if they are partly on grass. There is no difficulty whatever in the operation, which simply consists in cutting away the horn that overlaps the frog or the sole of the foot until the two parts are on a level. If dirt has got in between and is not dislodged by the paring, it should be scraped out, as it is liable to cause foot-rot.

CHAPTER XIV

BREEDING

GOATS in a wild or half-domesticated state breed but once a year. They pair usually from October to December, and bring forth their young from March to May, the period of gestation being twenty-one weeks. Domestication, however, shows its effect in this as in many other instances, and a goat that is well housed and fed on corn will sometimes breed twice in the year and at varying seasons. The best and most certain months for pairing are September, October, and November, as the kids will then make their appearance from February to April; thus by the time they are weaned the new grass has sprung up, affording a tender bite for their young teeth. But apart from this the earlier in the year they come into the world the better, as they have then all the spring and summer, when feed is best and most abundant, for their growth and development.

Although the best season for breeding is in the early part of spring, where several milch goats are kept it is not advantageous to have them all in full profit at this time, as the supply of milk would probably be in excess of the household requirements; whilst in winter the animals would be nearly dry, and then milk would be scarce. They should therefore be arranged to kid at varying periods; for instance, when three are the stock in hand, one should breed in April, and the others respectively in July and December if possible. These latter kiddings, during the months named, are, however, decidedly easier to recommend than to carry out. A goat may often be served in February, but the service is not so effectual in that month, and conception becomes less sure as the spring advances. For a goat to kid in December it must be mated in July, and it is quite a chance to get a goat in season in the hot months of the year. Indeed, when œstrum does occur then it is so transient that unless a male goat is running with the females it is difficult to get the animal served at the right moment; and even when service has been accomplished the goat does not always prove in kid.

The number of kids a goat brings forth at a birth varies from one to four. It usually happens that on the first occasion a single kid only is produced, but afterwards there are generally two and sometimes three. Four at a birth, although exceptional, is not very uncommon, and when it once happens is frequently repeated; indeed, I have noticed as a singular fact that, whatever be the number produced on the second occasion, the same will often be continued in subsequent litters, and not only this, but that the peculiarity descends from mother to daughter. A goat in my possession at one time, which had always three whenever she kidded, inherited this peculiarity from her parent. So large a litter, however, is by no means desirable, for when kids are numerous at a birth they are always small in proportion, and it is seldom that more than three can be properly reared by one goat, unless she is an extraordinary milker, three, indeed, being more than many can manage to suckle. A goat makes the best return in milk with its second or third lot of kids, continuing in its prime until six or seven years old. After that age the milk supply declines at each kidding, and the animal is seldom of much use for milking after nine years old, though I have known exceptions. The natural term of life is about twelve years, but I have heard of one living to eighteen and giving a pint of milk a day even at that age.

Age for Mating

In no domestic animal, I believe, is the instinct of reproduction so early developed as in the goat, its precocity being, indeed, almost incredible; thus I have known instances of kids being mated at three months old whilst still sucking their dams, and producing a live kid in due course. If a young she-goat is kept with the buck it is almost sure to come in season in September or October, whether born the previous

January or as late as May or June, so that it is much better to keep quite young stock separate from the male until they are old enough to be allowed to breed.

Before any attempt was made to improve the goat in size and milk production it was a common practice to let the doe have young when barely twelve months old, with the consequence that many of them were of diminutive size, having been thereby stunted in their growth. The quantity of milk obtained in such cases is, as a rule, very small, and the kids produced are often undersized. The earliest age at which a goatling[1] should have access to the male is fifteen months, but for exhibition purposes it is better to wait a month or two longer, to give time for a more complete development of the frame before the system has to undergo the drain upon it incidental to pregnancy and subsequent lactation. Although this may be at first more expensive, as the goat has to be kept longer, before making any return, it is probably the more profitable in the long run, the value of the animal being greatly enhanced, and the yield, I believe, increased.

As regards the male, it was thought at one time that he should not be less than two years old, but experience has shown this to be a mistake. Good stock have been got by bucks considerably under twelve months, but it is certain that after that age the male is quite capable of service with the best results.

Œstrum, or "Season"

A goat may come in season at any time after parturition, and until conception has taken place will continue doing so at certain intervals according to the time of year. From September to December, or as far on as January, this is generally repeated about every three weeks; thence till the end of March the intervals are longer, and from that month up to the close of August a goat may go the whole time without showing any signs of a disposition to breed. Much depends upon the system of feeding; where the animal is housed and fed chiefly on corn she may possibly come on several times during the summer months. The presence of a he-goat further encourages

[1] A she-goat over one year and under two years old.

this desire to propagate. The duration of the period is also dependent on the time of the year; in the autumn and early winter it will last three days, but in spring only two, and after that perhaps only for twenty-four hours or even less.

M. Crepin is of opinion that the she-goat is more disposed to breed when in an enervated condition. He has informed me that amongst the herds he has imported from the Alps, and which were brought to Paris by road, having been driven all the way, he has found that a considerable number came in season on their arrival, even in the summer months, being tired out by their long journey. The same authority states in his book that the goats of Malta and Murcia are more disposed to breed at unnatural seasons than those of the French Alps, the former having been accustomed to this by their owners, who have encouraged such a system from time immemorial and given special attention to their goats with this object.

Signs of Season

The "period" of the sexes may be recognised by an extraordinary restlessness on the part of the she-goat, a peculiar and continual bleating accompanied by rapid shaking of the tail. Her yield suddenly diminishes, and she loses her appetite, whilst there is also a considerable swelling of the vulva. If a male goat be in the neighbourhood, and the female be loose and able to escape, she will in many cases make straight for him though a considerable distance divide them.

Mating, or Pairing

When signs of œstrum are observed in the doe the services of the buck should be obtained without loss of time, and on their meeting the pair should be placed together in a stable by themselves. At the first interview it often happens that they commence fighting, though in their idea probably it is merely play; the tremendous blows, however, they inflict upon each other with their horns would lead anyone unaccustomed to their antics to suppose they were engaged in serious combat, suggesting the necessity of their separation at once. This need not be done, however, as after a time they come to a better understanding. It is not often that owners of stud-goats will allow their animals to be sent away from home, requiring, on the contrary, that the she-goat be brought to them. In such cases the she-goat is usually kept on the spot for a day or so, a charge of 4d. to 6d. per day being made for her keep. It is not necessary that the pair be left together for any length of time, provided the she-goat be quite in season, as impregnation generally takes place with the first service, though there is no harm in allowing a repetition in order to make quite sure.

I may here observe that when pairing takes place during the regular rutting season, *i.e.*, from September to January, it is nearly always successful, providing that the male animal is sufficiently vigorous.

Signs of Œstrum or "Season" in Pregnant Goats

It is generally supposed to be impossible for a goat to come in season any length of time after conception has taken place. Cases now and then occur, however, which lead to the view that this idea is not justified. An instance of this is recorded by the following case. A goat was mated at the end of August. About five or six weeks later she began to show all the signs of being in season, though at the same time refusing the attentions of the male. This was followed by the latter jumping the other goats, and the owner was led to suspect some disorder in the goat supposed to be pregnant. A veterinary surgeon was called in, who after examination pronounced the animal to be suffering from vaginitis (an affection of the vagina which is manifested by a thick mucous discharge from the vaginal membrane, and is sometimes very infectious). Injections were ordered and the owner of the goat was told that the animal would not "hold" until this was cured, the veterinary being of opinion that the goat was not in kid. Some weeks after this, when the period of gestation had nearly expired, a hard lump on the goat's side was noticed by the owner, and the udder was evidently springing. The following week a kid was born fully developed, proving that the goat was in kid when the signs of season or what resembled that condition were observed.

Choice of Stud-Goat

In a previous chapter I explained the points to be sought for in a good she-goat. I have now to make the reader acquainted with the necessary qualities of the male. Size is here of greater importance than in the female, therefore a fine well-developed animal should, if possible, be obtained.

A he-goat should have a small, neat head, with plenty of beard, neck short and thick and with abundance of hair. The horns, if any, may be large, but not too coarse and heavy. The chest should be broad and massive, the back long and straight, and the ribs well rounded, the tail being placed high up on the hind-quarters. The legs must be straight, thick, and strong, and well covered with hair on the thighs and buttocks. No better specimen of the type of animal I am describing could be found than that shown in "Monster" (Fig. 14), an unusually fine English stud-goat, winner of many prizes in the early days of goat-showing.

Importance of Pedigree

One of the most important conditions as regards the selection of breeding-stock is a good pedigree. When I say pedigree I do not mean merely a long list of ancestors without any qualifications, but a traceable descent from prize-winners on both sides, and in particular milking-prize winners, for the milking qualities are inherited as much from the sire's side as from the dam's. It is singular how this matter of pedigree has until comparatively recently been overlooked by goat-fanciers, and it is owing to the neglect of this important consideration that in the early days so few strains of goats could be depended on to transmit their characteristics to their offspring.

There is no difficulty in these days in selecting breeding-stock whose descent can be traced, as the Herd Book of the British Goat Society gives pedigrees as far back as 1875, whilst the Prize Record, which forms part of these volumes, supplies all the information necessary in regard to shows and the prizes won. Milking-prize winners have there a separate section, and all she-goats that have carried off prizes in

Fig. 14
OLD ENGLISH STUD-GOAT "MONSTER"

milking competitions are distinguished by a star placed against their names.

The "Prize Record," is a list of goats that have won prizes of any kind at Shows recognised by the Society. In early days considerable importance was attached to the winning of a prize, if only an award under inspection by a judge at Shows. In recent years this importance attaches chiefly to prizes won in milking competitions and more especially in such competitions as take into consideration quality as well as quantity of the milk yield. In 1909 a system was started to enable breeders to recognise at a glance the goats in a pedigree that had distinguished themselves as milkers at Shows. This was first accomplished by the affix of a star to the names of winners of milking prizes. Later these stars were for very good reasons not accorded merely to winners, but to goats that, in competing for milking prizes, had secured a certain number of points, whether winners or not. The female progeny of such a goat which in its turn won a similar honour was entitled to a double star and the third descendant under the same conditions a triple star and so on. These stars were given whether for quantity only or for quantity and quality combined. Then as quality competitions became more common it was decided to distinguish animals which won in milking competitions where quality was recognised from those which had only received points for a heavy yield, and these were indicated in print by the Q*. Male goats bred from star milkers and whose sires were also the produce of such milkers were distinguished from others by the printer's sign of a dagger (†) and were termed "dagger billies." This system has continued to the present day and is a great assistance in selecting breeding stock specially adapted for producing heavy yields.

Pregnancy

A goat may have all the appearance of being in young and yet be nothing of the kind, especially animals of advanced age. The fact may, however, be ascertained at six or eight weeks before the time for parturition arrives by pressing the fingers against the animal's flanks, when a hard lump may be detected, which is the head of the kid; when it is felt no doubt need be entertained as to the condition.

During the period of pregnancy, which varies between 147 and 152 days, a goat should be liberally fed, but as the time for the birth of the kids draws near—*i.e.*, within a few weeks—care must be taken not to over-feed the animal, as high condition at this time is liable to produce trouble during parturition, which may terminate fatally. To avoid this the rations should be changed and reduced in quantity, and only those of a laxative nature supplied; very little corn must be given at this time, the diet being restricted chiefly to grass or hay and garden produce. Mangolds are frequently fed to goats that are in-kid, but some breeders are of opinion that these are apt to cause abortion if given in a frosted condition. A bran-mash may be advantageously given every other morning, as it keeps the bowels open.

Goats as a rule drop their kids easily, and in nine cases out of ten without accident, such a thing as a goat dying at the event, even when the young are born dead, being quite the exception. It is somewhat surprising the amount of butts and blows these animals are capable of receiving from their fellows while pregnant without experiencing any ill effects. It would almost appear as though Nature had specially provided against all such casualties. Nevertheless, it is advisable to take extra care of goats in-kid, to prevent all chance of abortion, for if one goat slips her kids through some unforeseen circumstance, others in the same stable are very apt to follow her example without any apparent reason. They should, therefore, be driven or led gently, and not allowed to take violent exercise, as much exertion at this period is prejudicial. For the same reason no dogs or other animals likely to cause a fright should be permitted to approach. It is very necessary, however, that gentle exercise be given to goats that are pregnant, as parturition is generally more easy in consequence.

Goats that live on a common are generally hardier than such as are kept constantly in a stable. The former will often drop their kids under a furze-bush, and, after licking them over, will partly cover them with ferns and go on browsing, returning at intervals to suckle them. One that has been more delicately reared, however, requires more careful treatment.

As the day of her expected accouchement draws near, a separate place should, if possible, be prepared for the dam, where she may move about at will, as it is important that she be not fastened in any way.

Signs of approaching parturition are shown by the enlargement of the vulva ; the goat becomes restless, and lies first in one place and then another, frequently changing position ; she also bleats a great deal, and her udder becomes rapidly harder, being replete with milk. When this is noticed she should not be taken out to graze, but kept in the stable, where a bed of clean straw, a bucket of tepid water, and some nice sweet hay should be provided for her.

Treatment after Kidding

While giving birth to the kids a goat is best left to herself, as these animals rarely require any assistance. As soon as it is over, which may not be the case, if there be several, for an hour or more (thirty or forty minutes elapsing in some instances between the birth of kids), the mother should have a drink of oatmeal-water, made by pouring boiling water on a handful or two of coarse Scotch oatmeal, and allowing it to stand until lukewarm ; on no account should quite cold water be given at such times. After this drink, which is the first thing the goat looks for, a bran-mash may be offered her. The mash is made by pouring boiling water over a quart or so of bran until thoroughly wet without being too sloppy.

The first act of the newly-born kids upon gaining their legs, which they do, if healthy and strong, almost as soon as they are into the world, is to make for their mother's teats. These they are sometimes slow to find; but that is of no consequence, as they do not actually require nourishment for several hours. If the udder be full of milk, however, a *portion* should be drawn away, as the great weight and pressure will cause the mother pain, and if the milk be not removed it is apt to form clots and produce inflammation ; when this happens, as will be noticed by the great heat of the bag, it should be fomented with hot water, rubbed over with vaseline or goose-grease, and well worked in the hand. As soon as the milk comes freely the kids should be put to the mother, and if they do not then suck properly, the teats must be placed in their mouths ; once they have drawn a few drops, they rapidly improve the occasion, and do not often require a second lesson.

For the first three days the milk has a highly yellow appearance, and, containing as it does a considerable proportion of mucus in its composition, is not well adapted for domestic use. Upon the kids, however, it acts as an aperient, and is very beneficial. The milk assumes its ordinary character in three or four days, and this may be known to be the case when it froths up on being drawn.

During cold weather it is better to keep the kids shut up with their dam in the goat-house ; but should the days be fine and dry, especially in spring and summer, they may be all put out the third day. With good milkers it often happens that the kids, if only two in number, do not take all the milk that the mother yields ; in which case it is advisable to draw off a portion daily while this lasts. It is a mistake, however, to draw more than a small quantity of milk at a time, and preferable to take none at all if the kids can consume it. The goat requires careful watching the first day or two to see that both teats are properly drawn. When there is only one kid, and the goat is a heavy milker with unusually large teats, the kid will often suck one teat only, leaving the other untouched. The result is that the neglected teat gets very distended, and, if allowed to continue in such a state, becomes corded by the milk clotting inside, and is rendered useless ever after. Even when this does not happen one teat will become larger than the other, which is objectionable and would tell against a goat competing for a prize. The simple prevention in such cases is to milk the untouched teat regularly night and morning until the kid takes to it, which it will generally do after a few weeks.

One of the troubles which the goat keeper not infrequently has to contend with is the condition known as barrenness which not having been dealt with in earlier editions of this work, must find a place in these pages. The subject has been admirably treated in more than one of the Year Books of the British Goat Society by Mr. H. Stainton, F.R.C.V.S. who has perhaps given more attention to the study of goat diseases and contributed more instructive articles upon it than any other member of his profession. From one of these articles I venture to quote rather freely as the subject is an essentially technical one.

Sterility has many causes but there are three factors which we are told militate against productiveness: "1. Unnatural conditions. 2. In breeding. 3. Intensive breeding, such as that carried out for the purpose of heavy milk production." This last describes a practice which we all know to be very generally indulged in at the present day and often with dire results for it is forcing nature and this can seldom be carried on for any length of time with impunity. "There is disturbance in endocrine secretion, there is faulty ovarian activity and in short deficiency in the reproductive powers."

Sterility may occur in either the male or the female, in the former due generally to some malformation of the generative organs. In the female the causes of failure to reproduce are very numerous some of which, as in males, are according to the authority quoted, anatomical, while others are either pathological or nervous and temperamental. Most experienced breeders are familiar with cases where the she goat comes into season continuously, repeated matings being unsuccessful. This abnormal condition is probably caused by a persistent ovarian cyst and is known as "nymphomania." The treatment recommended is isolation, saline purgatives and service towards the end of œstrum. The lethargic fat subject, usually an overfed pet, is also likely to fail to breed and for this, exercise, limited concentrated diet and a mild sexual tonic is advised. Some goats show failure through an anæmic and unthrifty condition which may be "due to parasites, worms, low feeding, etc. The cause must first be removed and the subject then well fed." These three are nervous and temperamental cases and treatment is said to be usually effective. Other illustrations of deficiency in the reproductive powers are given as follows:

"1. A goat breeds normally up to, say, three years of age. At this period she may have a difficult kidding—perhaps accompanied by retention of some afterbirth. This clears up but subsequent matings are fruitless over a long period."

"2. The same as above—but there may be periodic hæmorrhages. Also at intervals the goat shows evidence of some discomfort in the region of the loins by characteristic humped-up attitude and by looking round."

"3. The young animal comes into season and is mated. No result—but in the course of a week or so apparently is in season again and remains so, spasmodically, more or less indefinitely. In either case mating is useless."

There is again the simple case of vaginitis or catarrh of the vagina where (we read) "a goat of proved breeding power has obstinate vaginal discharge and although normal in all other respects fails to take after repeated service." This is quite common.

Another or probably quite unsuspected cause is due to fretting and nervous depression to which the petted animal in particular is very subject. "It is sent away for service, is mated apparently successfully but may be kept at stud until the next œstrum to make certain. She returns home a physical wreck and is not in kid."

The more serious cases, it would seem, however, are due to ovarian derangement the result often of uterine trouble as instanced in No. 1 above.

Among extreme cases are the well-known one among goats where the animal proves to be "a freak of nature" representing a combination of the two sexes. These are known as Hermaphrodites and are invariably, and, I believe, incurably infertile.

Although, as we are told "many of the conditions operating against conception are easily understood, diagnosed and treated," speaking from personal experience I should recommend doing away with the goat that proves barren after thorough and repeated tests. Where, however, the animal is particularly valuable by virtue of its milking qualities or its milking parentage it may be worth while to submit the case to a qualified veterinary surgeon.

CHAPTER XV

REARING KIDS

As a general rule it does not pay to rear kids, and certainly not the males, unless they are from superior stock and likely from their parentage to breed valuable milkers. A short calculation shows this very clearly: Suppose a goat has two kids, and gives 3 pints a day, which is the least she should yield to bring up a couple properly, at the end of six weeks they will have taken 126 pints, with an additional 21 by the time they are weaned—in all 73½ quarts. Reckon this at the lowest possible value, 4d. per quart, the price of cows' milk in the country, and you get 12s. 3d. as the cost of each kid; or, supposing the goat to have but one kid, and be giving a quart daily under the same circumstances, that one kid would cost 16s. 4d. Now, considering that, if a female and a common specimen, it will only realise from 7s. 6d. to 10s., or, if a male of no better quality, be hard to sell at 6s. or 7s., it becomes plainly evident that this system, to say the least, is not a profitable one. When it is desired to establish a herd of a good milking strain, and a high stud fee has been paid, the case is of course different. One does not then look so much at the first outlay, in the hope of obtaining in the future some valuable stock. It is better to kill off the male kids, unless they are superior animals, at a fortnight or three weeks old; and have them dressed for the table. They are then delicious eating.

Weaning

With people who want the milk from the goat it is a common practice to commence weaning the kids at about six weeks old, but if fine stock are required they should be allowed to suck as long again, this being the great secret for gaining size. In any case the weaning process should be a gradual one, the kids being at first separated from the dam during the day, fed on hay, bran, and crushed oats, or, if pastured, put to feed where they will find some young and tender grass and herbs. They will eat such food very readily at six weeks old, as they begin to make use of their teeth when only a fortnight old, and sometimes even earlier if the milk supply is scanty. When the separation can be made to take place at a distance from the owner's residence he will find it beneficial, for his own sake, as the frightful noise these youngsters and their fond parent set up when they discover themselves parted is enough to make a nervous and impetuous individual put an end to the existence of the lot in the most summary manner possible. When separated at a sufficient distance to prevent their hearing each other's cries, however, they soon get accustomed to their position. After six weeks, before the goats and kids are placed together at night, if the milk is much wanted the goat may be relieved of a portion of her milk, allowing the kids afterwards to take what remains.

Rearing by hand

There is another way of rearing kids, and although it seems less natural it has some advantages over the ordinary one, especially for the goat. This consists in bringing up on the bottle, using for the purpose an ordinary infant's feeding-bottle. Most dairy farmers at the present day prefer to separate the calf from the cow either immediately after birth or within a couple of days, and to bring up the young stock by hand.

Different opinions exist as to the best time to separate the young from the dam, some holding, and with perfect truth, that the latter frets much less at the loss of her offspring if removed directly after its birth than when her affection has become intensified by suckling it. On the other hand, it is argued that the sucking of the young has a beneficial effect upon the goat and her yield of milk—"drawing down" the udder, as it is expressed, more freely. For my own part, I prefer to let the kid suck for three days, or at most four, so as to let it get the "beestings" or first milk, so specially adapted to its requirements at this stage, and then to remove it and bring up by hand. The earlier this is done the easier will the kid take to an artificial teat, but during the first three days there is but little difference in this respect. As to the fretting of the goat for her young, this soon subsides if the latter be removed sufficiently far away to prevent her from hearing its cries. When this cannot be accomplished, the plan of separating at birth is best.

Another way is to teach the kid to drink from a pail by introducing one hand into the milk and projecting the second finger a little above the level of the fluid as something at first to suck at, this being after a time withdrawn ; the other hand is used to keep down the head of the kid, but it is important not to insert the mouth of the animal so far down into the milk as to cover the nostrils. Strange to say, most of those breeders who have adopted the hand-rearing system practise this method of feeding, probably on account of the little trouble it gives after the kids have once acquired the knack. I cannot personally speak in its favour, however. First, I have always found much trouble in making the animals learn to drink at this early stage of their existence, and when learnt it is, after all, injurious for them to take their food in this way. It has this serious objection, that the milk, being gulped down in large draughts, is introduced into the stomach without a due admixture of saliva, which performs so important a part in its digestion, and causes masses of coagulated curd to accumulate in the stomach, producing diarrhœa, which not uncommonly ends fatally. In sucking, the saliva mixes with the milk in the mouth, and the whole goes down in a more gradual and regular manner ; but, should the orifice of an artificial teat become much worn and enlarged, or the end perforated by the teeth of the kid, the same result will be inevitable.

Goats' milk must constitute the sole food for the first week, after which fresh cows' milk may be added in gradually increasing quantities until given alone at the end of a fortnight ; a week later some good skim-milk scalded and cooled to ordinary milk temperature may be mixed with the new milk in the proportion of one-third at first, and a half a few days later. At this stage some linseed-tea, peameal broth, or oatmeal gruel may be introduced into the milk, in order to increase its feeding properties, the quantity being augmented slightly each day, until at eight weeks old the kid may have the artificial food alone. Linseed-tea is made by boiling half a pound of the seed in five and a half pints of water; it takes the place in some degree of the cream that has been abstracted from the milk. Peameal broth is not boiled, but made into brose by pouring boiling water on and stirring until fine; this jellies on cooling, when it must be intimately mixed with the milk, so that the whole is free from lumps. The composition I prefer, however, is what is known as Lancaster Calf Meal, made by Messrs. Bibby and Sons, of Liverpool, which is much preferred by kids to any of the other artificial foods above mentioned.

Kids reared by hand in this way require regular and careful feeding, beginning with a pint daily for the first three days or a week, according to the appetites of the animals, and increasing the quantity as they are able to take more; they will be far healthier, however, and thrive better with a little under-feeding than with too much. One over-dose may cause scouring, which will throw them back for days. Four meals a day should be given for the first fortnight or so, and three afterwards. In six weeks they will be eating hay and grass, and may also have oats to nibble; but the artificial feeding should continue as long as they are willing to take it.

By being accustomed to sloppy food in this manner for some time after weaning they will take it ever afterwards, and when they become mature goats and have kids themselves such feeding has a great effect on the production of milk. Besides this, plenty of liquid food encourages a big stomach, which, with a good digestion, is a marked feature in abundant milkers.

Treatment of Kids after Weaning

The treatment of kids after they are weaned must depend on the purpose they are intended to serve. Those for future stock need plenty of exercise and good feeding, as great a variety of food being given them as possible. The fact that some goats readily devour anything, whilst others are fastidious and care only for corn and hay or grass, refusing even some kinds of the former, is doubtless to be accounted for to a great extent by the way they have been brought up. For a goat to be a good milker it should be a great eater and drinker, and should be so constituted by nature that the food it consumes is transformed into milk rather than into meat or fat; hence it is that heavy milkers are usually thin although well fed. A goat, however, is like a human being, inasmuch as it becomes satiated after a time with one kind of food, but will continue to eat with avidity when something fresh is offered to it. Thus it is important that young growing goats should be early accustomed to such things as bran-mashes, grains, pulped roots, linseed cake, etc.

For the first six months or thereabouts, kids should be fed four times daily; oats, crushed or whole, and bran (which is specially good to make bone and muscle) should be the staple food, with grass and the articles mentioned above, or some of them, supplied occasionally for a change. In introducing any new food, if disliked at first it should be mixed in gradually increasing quantities with something that is relished. Linseed cake, for instance, should be broken up very small, and put in with the oats and bran; pulped roots also may be mixed with bran or dusted over with meal. The quantity given cannot be specified, as so much depends on the individual appetite, but more food should never be placed before the animals than they are likely to consume.

After the age above mentioned, should the season be spring or summer, more grass may be allowed and less corn given. Green food causes the stomach to enlarge and develops the digestive organs; but it must not be supplied too abundantly in spring, nor too suddenly at any time, for fear of causing relaxation of the bowels.

Scours in Kids

Kids are always more or less apt to be attacked by scours, more especially if too much food is given at a time. This may be stopped (though it should not be done too suddenly) by administering a draught, the recipe for which is given under Diarrhœa in the chapter on Diseases. The excretion, therefore, should be carefully watched. On the first sign of looseness give a mess of baked flour boiled up with milk or milk and water, not too much at once, two or three times a day. Those who have no nurseries may be told that this is made by putting dry flour in an earthen jar in the oven, and shaking it occasionally till it is a deepish whity-brown. Rice milk is also good, or, for severer cases, bean-meal gruel. All should be given cold. If the diarrhœa has been severe, so that the little creature has moped, drooped, and dwindled, further danger will arise when the return of the excretions to their natural form shows that the purging is over. Obviously much green food must then be avoided. Nor should the kid be allowed to eat heartily of dry stuff or grain if it can get at what older goats have, under the notion that it wants "feeding up." The stomach after diarrhœa is excessively delicate; if it is overloaded the kid will cease to eat, become distended, and die, the digestive system having stopped altogether. A tiny allowance of grain—say half a gill of bruised oats at a time—twice a day, or a little broken biscuit or bread, with hay, is quite sufficient after the binding drinks have been left off. Green food should be gradually reintroduced, the great principle of feeding after diarrhœa being a little at a time. Remember also to keep from damp and cold night and day. If the droppings are black, hardish, and pointed at the ends, instead of round and fairly moist, as they ought to be, confinement of the bowels and indigestion are indicated. A very young kid, however, on spring grass and linseed-tea is not likely to be constipated.

Exercise for Kids

It conduces greatly to the well-being of a kid if it can be permitted to have free exercise, though on grounds that are only protected by hedges this can seldom be allowed, for these little animals, in whom the mischievous propensity of their kind is inherent, soon find out where their favourite food is located, and either by jumping—if the obstacle be a low wall—or insinuating their small bodies through the slightest opening, should it be a fence, rapidly make their way thither. Tethering of course may be adopted, but it should be avoided if possible for goats at this early stage of their existence. When practised, however, care must be taken that the tether be fixed free of all obstacles, otherwise in its gambols a kid is very likely to get the rope or chain caught, and strangulation may ensue. It is better never to use a cord in keeping a kid within bounds, but to employ a light chain well supplied with swivels, and to see that it works freely with the pin. If a stake is used it should never be placed above ground, but driven in so that the top comes level with the soil. When this is neglected the little animal is sure to wind the tether round the stick until it is so close up that it is no longer able to move. A small head-stall, carefully made to fit, is better by far than a collar.

Kids that can be allowed an enclosed yard in which to exercise at will may with advantage be supplied with some means for indulging their natural propensity for climbing and jumping. This can be very simply managed by driving posts into the ground a few feet apart, and nailing to them cross pieces of wood such as round or half-round poles—preferably fir poles—with the bark on. A board must be placed obliquely at one or both ends to enable the kids to climb up to this rough kind of gymnasium, the height of which should be about 3ft. from the ground.

Few would believe the amusement that two or three kids, when provided with such an arrangement as I have described, will furnish an observer. Indeed, I may admit that I have frequently spent an hour at a time in fits of laughter witnessing the gambols of these amusing little creatures. They soon learn to make use of their gymnasium, and after a few attempts will mount the structure with a skip and a bound ; each starting from opposite ends and meeting in the centre will fight for "king of the castle" like two schoolboys, trying which can push the other over, one of the two and sometimes both being butted off. Even if the horizontal pieces be round poles peeled of the bark and slippery like a scaffold pole, it will not prevent these sure-footed little quadrupeds from walking across, though the feat is performed with less rapidity. Exercise of this kind, whilst highly entertaining to the spectator, is most beneficial to the kid ; it develops the muscles, strengthens the joints, and by promoting circulation encourages health, besides in cold weather keeping the animal warm.

Weights and Measurements of Kids

A kid that promises well is usually unsymmetrical in shape, its legs being long and thick and out of proportion to its body. Showing this feature with well-sprung ribs and a good development of chest, such an animal may be expected to make a fine goat.

Breeders are often desirous of getting some idea of the merits of their young stock in the matter of growth and development, and I am often appealed to for information on this point. I have at various times taken the weights and measurements of prize goats and kids at shows, and have received reliable data from well-known breeders.

Mr. Rothwell-Jackson's Lancashire Witch, a famous prize-winner, weighed the day after birth 6¾lb., at four weeks old 16½lb., at eight weeks 25lb., at twelve weeks 35½lb., at three calendar months she turned the scale at 40lb., and she added 10lb. to her weight every month after until at six months she weighed 70lb. These weights were extremely good, as many she-goats fully grown only turn the scale at 84lb.; and yet Lancashire Witch was by no means a large goat in the adult stage. Another, and a much bigger animal as a goat, at four weeks weighed 18lb., increasing to 26lb. three weeks later. At two months she registered 30lb., at three months 41½lb., at four months less ten days she weighed 48lb., and at five months 59lb. At ten and a half months this goat measured 26in. at shoulder and weighed 83½lb.

Lorna Doone, out of Lollo by Garnet, however, beat both these kids. Her weights were at five days old 8lb. 9½oz., at eight days 9lb. 11oz., at fifteen days 12lb. 2oz., at twenty-two days 14lb. 5oz., at twenty-nine days 17lb. 8oz., at thirty-six days 21lb. 3 oz., and at two months 32lb. 13oz. Yet this was never a very large goat, for when over two years old she only stood 28½in. The Baroness Burdett-Coutts's Gertie highly celebrated as a first-prize winner, which became one of the heaviest goats exhibited, at seven and three-quarter months measured in height 27in., was 32in. in girth behind the shoulders, and weighed 70½lb. All these were Anglo-Nubians.

Coming now to half-bred Toggenburgs, Bridie, a she-kid then in my stable, out of Daffy by Sedgemere Paris, weighed at one week old 14lb., at two weeks 17lb., three weeks 20lb., four weeks 24lb., five weeks 27lb., six weeks 30lb., and so on at the rate of from 3lb. to 3½lb. a week for some weeks after.

Of male kids I find the following recorded. Battenberg, a first-prize winner as a kid, at five and three-quarter months weighed 59lb.; and McNab, at seven and three-quarter months, 74lb. Hornblende, a first-prize winner as a kid, weighed 83lb. at eight and three-quarter months, and its fellow Midnight, a second-prize winner, at about the same age, 66lb. More recently Mr. Sam Woodiwiss's Sedgemere Sergius at seven months weighed 69lb., and at nine months 87lb. Peter Pan, a second-prize Anglo-Nubian kid, at three and three-quarter months was 27in. in height, 31in. in girth under the shoulder, and weighed 60lb.; whilst Polidore, a Toggenburg, at the same age, though only 26½in. in height, was reported by his owner to weigh 72lb.

Breeders will find it useful to weigh their young stock periodically and to compare the weights given with some of those here mentioned. The late Mr. Rothwell-Jackson, who weighed his stock most systematically and was a constant winner of prizes, arrived at the following conclusion, as expressed in one of his letters to me: "My weights show one rather important point, which is that, supposing one does not care to rear more than one kid out of a litter, it by no means follows that it is wise to keep the largest, as sometimes the smallest will prove the best; but in every case, according to my records, if the smaller animal is going to be the better eventually, it will have overtaken and passed the other before a month has elapsed, so that unless one kid at birth should be very superior indeed to the other, it is safest to let them stay with the dam for at least three or four weeks, to form a more correct idea as to which is likely to be the better."

CHAPTER XVI

MILKING

MILKING is not so easy an operation as at first sight may appear, and it is only properly learnt after some little practice. There are two ways in which it may be performed—the one being termed "nievling" and the other "stripping." The former is that generally practised by milkmen and dairymaids, and is certainly the better of the two. It is effected by firmly grasping the teats, one in each hand, and gently but rather quickly forcing them down, at the same time closing the fingers upon them, the pressure commencing with the first and terminating with the little finger. As soon as the stream thus caused to flow has ceased, the grasp is slightly relaxed, and the hands are sprung quickly upwards, when a fresh supply rushes from the udder, and is again squeezed out as before, the streams from each teat following so closely in succession as to sound like one continuous flow.

Stripping consists in taking hold of the teat at the part nearest the udder between the fore-finger and thumb, which are slid down its whole length, exerting at the same time some considerable pressure, thus causing the milk to flow in a forcible stream. This process is continued until the udder is completely emptied.

Of these two modes of milking, nievling is, as I have remarked, the better to practise, being done with greater comfort to the animal—resembling more the action of a kid sucking—and taking up less time. With some goats, however, it cannot well be accomplished, the teats being so small as to afford no permanent grasp for the hand, and therefore stripping must be performed. The latter operation, however, is always resorted to at the termination of the milking to extract the last drops, it being easier and more effective for that purpose than nievling. Many persons also commence with it, and continue until the milk flows freely, when the other process is resorted to.

But milking cannot well be learnt from any written instructions. I would therefore advise the novice to get his milkman or some other proficient person to come once or twice a day for a week or ten days, and to watch him perform the operation, proceeding afterwards with it himself under the man's supervision until he has so far mastered the art as to be able to milk "clean," as it is termed—*i.e.*, to draw every drop from the udder. He must not be disheartened if at first the milk comes very slowly and only after great exertion on his part ; this difficulty will soon disappear as he gets more accustomed to the work. He should practise on a tame, quiet animal with a gentle disposition, as some are very troublesome with strangers, holding back their milk for a considerable time, and then only giving it in a fine thread-like stream, which means a long while to complete the process. Goats with good-sized teats are more easy to milk than those with small ones, and should therefore, if possible, be procured to learn on, though the freedom of the flow is not governed by the size of the teat.

Two important points are to be borne in mind in milking—viz., to milk quickly and to milk "clean." The former is necessary for two reasons : first, more milk is said to be obtained when rapidly drawn ; and, secondly, a goat will often get impatient when the operation is performed lazily, and will suddenly kick up her hind legs and probably overturn the pail. The importance of milking clean cannot be over-estimated, as neglecting it results in the animal soon going dry. Many a beginner will rise from his task under the impression that he has obtained all, when an experienced hand will afterwards, to his surprise, draw perhaps another quarter of a pint or more. This extra quantity, when all has been apparently obtained, may generally be secured by imitating with the hand the action of the kid when sucking, *i.e.*, sharply pushing up the udder with the closed fist, the teat being at the same time retained in the grasp. It is surprising the effect this has in inducing a fresh flow. The reason why it is so necessary to get at each milking every drop of liquid that the udder contains is that when any is left behind it is absorbed back into the system, serving to indicate that a proportionately less amount is required during the following interval, the consequence being that the animal in a short time becomes dry.

Where several goats are kept it is a good plan to milk all the "strippings" into another pail, using them in tea and coffee, and keeping the other portions for cooking purposes. The reason why the "strippings" or "afterings" are so much richer than the previously drawn portions is that the fatty particles rise to the top of the milk as contained in the udder in the same way as when afterwards placed in open vessels, and as the lower part comes out first it follows that the greatest amount of cream only makes its appearance at the termination ; hence the advantage of drawing every drop that the udder contains.

Regularity in Milking

Regularity in the hours of milking is a point of great importance if an even and constant supply is desired. Indeed, with the exception of food and water, nothing influences the yield more than this. A delay of a couple of hours, which may cause an extra quarter of a pint to be drawn, will show its effect at the next milking by the loss of half a pint at least, whilst an alteration of three or four hours may affect the supply for a day or two. A goat that gives three quarts or more should, for the first three months after kidding, or as long as in full profit, be milked three times a day—at seven o'clock in the morning, shortly after noon, and at seven in the evening. Afterwards, two milkings daily—about seven a.m. and seven p.m.—will be sufficient. This should be kept up until only a pint is given in the twenty-four hours, which need only be drawn once a day, preferably every morning.

Cleanliness and other Conditions

Cleanliness is another matter of importance, at least in my opinion. Who has not noticed with disgust the milking of cows on a dairy farm, where the milkman goes to work with hands begrimed with dirt, on which to facilitate the process he first expectorates, and as the liquid flows squirts a few drops into his hands or dips them into the pail? Whichever he does, the result is generally the same at the termination ; the last drops that are stripped from the udder combine with the black on the fingers, forming a muddy compound, which, when the pail is emptied, leaves a sediment of dirt at the bottom. I do not, however, mean to insult my readers by supposing they would milk their goats in this fashion with dirty hands ; it is to the teats I wish to draw their attention. These when the animal has been lying all night on soiled litter, or during the day on muddy ground, are sure to be more or less contaminated with excremental matter, and, if not attended to, the result must be very much the same as in the case just alluded to ; it is therefore advisable previous to each milking to go over the teats with a cloth or sponge and *warm* water until clean, and to finish by wiping them with a dry cloth. The milk seems to flow much more easily after this ablution, and no fear need then be apprehended of a deposit showing in the pail or of a disagreeable taste in the milk. Indeed, I prefer to go even further than this, and add a small quantity of some harmless disinfectant to the water, such as a trifle of permanganate of potash, which thoroughly purifies the teats and removes any unpleasant smell. This may appear fastidious to some folk, but the rapidity with which the fluid changes colour shows that the disinfectant is by no means unnecessary. *When teat and hands are clean*, a few drops of milk may be advantageously squeezed into the palms, and no unpleasant consequences ensue ; it acts as a lubricant, and makes the operation easier not only for the milker, but for the milked. Otherwise dry milking from the first is preferable. Goats should not be milked with very cold hands any more than with dirty ones ; some of these animals are very sensitive in this respect and do not "let down" their milk so freely as when the hands are warm. Another point to observe is not to use soap or soda in washing the teats.

Position, etc., when Milking

Milking may be performed either from behind the goat or at the side. In Malta and some other places abroad where the goats have generally long rather than spherical udders, the former is the plan adopted. It can only be practised properly, however, with goats that have such udders, but even then it does not commend itself to me as being so cleanly or so convenient as side-milking. Most milkers kneel down to the work; but in this position, when the operation is protracted, it becomes an exceedingly tedious, not to say painful, process. It is far better to sit on a low stool and have the goat on a raised bench about 2ft. from the ground. This is the plan I have adopted for many years, and I find no difficulty whatever in getting the animals to jump on to the platform after the first two or three lessons; in fact, they get so accustomed to the arrangement that at the usual hours, as soon as they are released from their stalls, they rush to the milking-bench, and jump upon it without the least trouble. Of course the secret lies in the fact that their corn is there waiting for them, and they know it.

When this plan is adopted it is necessary to feed and milk at the same time, and it serves two purposes—if the goats are milked regularly they get fed regularly, and also stand more quietly during the operation when their attention is engaged on their food. The objection to this system is that if a goat is not giving much milk the milking is finished before the feeding, and one has then to remove the contents to the stall for the animal to complete its meal there. When a goat has been once accustomed to the milking-bench and stands quietly during the operation, the feeding may be accomplished first, and she will then just chew the cud whilst her milk is being drawn from her.

Some goats are, however, a great trouble to milk, especially at first, as I have before remarked; indeed, I had one that was quite six months before she would stand quietly. I tried various modes of tying her up, but without avail; if she could go neither forward nor backward she would manage to rear up on her hind legs and then lie down altogether, rendering milking simply impossible. I accordingly devised an apparatus which succeeded so well that I give it here for the benefit of my readers.

At the head of the stable,[1] in front of the manger, I fasten

[1] I have since erected the milking-stalls in a separate part of the goat-house in conjunction with the milking-bench described above. (See plan of goat-house on previous page.)

on each side of the wall a batten (*a a*, Fig. 15) 5ft. long, on which I nail two pieces of wood (*b b*) to form guides or grooves. In these guides are run three boards 9in. wide and 1in. thick, the lowest within 3in. of the floor, and supported in its place by means of pieces of wood (*c c*) nailed horizontally across the bottom. The object of this space is merely to be able to rake from under the feeding-box any waste food or refuse that may have fallen and accumulated there. Two similar pieces are nailed at the upper extremity of the guides to prevent the top board from sliding out. In the view here given only

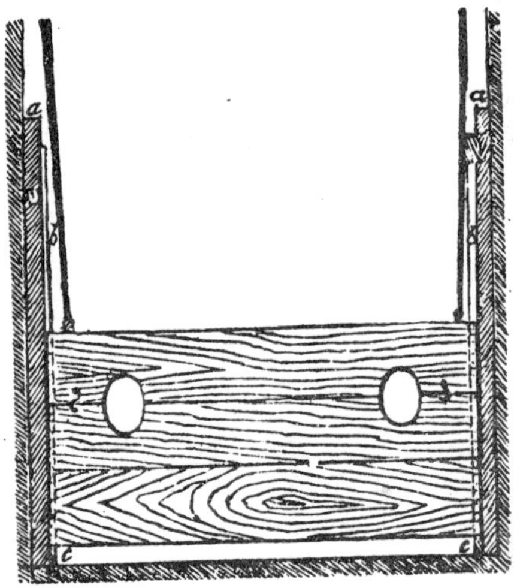

Fig. 15
"GUILLOTINE" MILKING-STALLS

one guide is shown on each side, to allow the working of the board to be better seen.

At the top of the middle plank, and 10in. from each end, a piece is cut, or rather sawn, out, leaving a semi-oval opening, the one above having at the bottom a corresponding piece removed, so that when the edges meet an oval hole is formed, measuring 7in. from top to bottom, and 6in. across. The object of this, as may be readily guessed, is to hold the animal's neck, it being easily effected by sliding up the top board, and as soon as the head is inserted over the lower one letting it drop, securing the two boards together with a couple of hooks and eyes. The upper plank is arranged by means of two weights and some sash-line run on pulleys to slide up and down like a window without any trouble or exertion.

The dimensions I have given of the holes and their distance from the ground are in accordance with the sizes of my goats' necks and their heights. They may, therefore, be made larger or smaller to accommodate the animals for which the apparatus is intended. In any case they must not be too large or a hornless goat will withdraw its head.

When the milking-bench is used these stocks require to be somewhat modified in construction; they are then narrower in width, one opening only being required; the sliding-board works all the better for this, as it is less apt to catch at the sides. It is scarcely necessary to add that the stocks are raised to the required elevation, according to the height of the bench.

It will be found advantageous to have a bench and stocks fitted up in the goat-house, and to use them with all goats, whether restive or not; the animals are then accustomed to their use, and if at any time they become restless at milking they are at once subject to control.

It must not be supposed that all goats are the same trouble to milk as the one I have described. On the contrary, as a general rule they are, after the first few times, quiet and tractable, submitting to the operation with apparent pleasure. Nor is this to be wondered at, considering the hard and distended state of the udder of a good goat previous to its being milked, and the symptoms of uneasiness the animal exhibits if the milking hour is delayed. When the milk is drawn for the first or second time after kidding, the animals are apt to be rather fidgety, but this will soon be got over by kindness and patience.

Here let me remark that patience and an even temper are two essential qualities for anyone to possess who has much to do with milch stock. It is said that cows will give more milk to a gentle, kind-tempered milkmaid than to one with rough and impatient manners, and that they prefer to be milked by a woman. However troublesome and unsteady your goats may be, keep your temper as much as possible, patting and coaxing rather than punishing by blows, for much of the latter treatment is not unlikely to result in an empty pail, with the subsequent effect of the animal going dry altogether. I have certainly in some instances found a sharp stroke with a switch the only effectual cure for the habit contracted by some goats of lying down as soon as their teats are touched, but measures of violence should only be adopted as a last resort.

Dairy Utensils

A few necessary utensils will be required for use in the "goat dairy." These are a milkpail, a strainer, a cream-skimmer, and shallow pans for containing the milk when set for cream, unless a small hand separator be used. Round wooden bowls used to be in vogue for milking, but these are now superseded by small galvanised or, what are preferable, enamelled iron pails, as the latter are easier to keep clean and sweet ; moreover, a pail is less easily over-turned should the goat place her hoof upon its edge. After the milk is drawn it should be strained through a fine hair-sieve. The object of straining is not so much to get rid of dirt, which, as I have shown, ought not to be present, as to retain in the sieve the hairs from the udder, which in some seasons come off in large quantities.

Every utensil should be scalded with boiling water as soon as used, and afterwards put out to dry and purify in the open air. An occasional soaking in soda-and-water is further beneficial in case any milk should not have been thoroughly removed, as when milk turns sour and decomposes it gives an objectionable flavour to any fresh milk that the vessel afterwards contains. It is generally this want of cleanliness that gives goats' milk its peculiarly nasty taste when drunk abroad, and prejudices people against it ever after.

The Milk Yield of a Goat

There is a vast difference in the yields of milk of different goats, some giving so small a quantity that the value of the total supply during the whole period of lactation is hardly equal to the cost of food. In such a case, should this deficiency of supply continue a second season the animal had better be disposed of. It must be remembered, however, that not only is there a great variation in the productiveness of different goats, but that the same animal may give considerably more milk one year than another, though this is the exception rather than the rule. It must also be borne in mind that at the first kidding the yield is usually much less than in the second or subsequent years, especially if the goat is under eighteen months old when the first kids are born.

For a goat to be a good milker she should supply at least three quarts a day, and continue this for three months or more. One that gives only half this quantity when the kids are first weaned is not worth keeping.

A goat may be a good milker, however, in two ways: by the quantity yielded at the full flush, and by the duration of her milking period. A goat is more valuable as a milk supply that gives two quarts a day at first and continues milking fairly well until within a month of the birth of its next kids than is one that gives three quarts after parturition and only remains in milk for six months.

A great advance has been made in the productive quality of goats in the last few years. As will be noticed in the preceding pages, a gallon milker was regarded at one time as next to an impossibility and now we have goats that have very nearly doubled this. Mrs. Abbey, of Didgemere Hall, Essex, produced an exhibit, Didgemere Dream, at a Show at Yarmouth on 15th June, 1926, which at one milking gave one pint over the gallon in fourteen hours and reached 18lb. 7oz. in twenty-four hours. Gallon milkers in these days are quite common even at milking competitions.

Influence of Water on the Yield

The amount that a goat drinks has a direct bearing on the yield of milk, both in quantity and quality. It might naturally be supposed that the quality would be reduced as the quantity was increased, the addition being merely that of water, but from certain experiments that have been made by Daniel and repeated with additional precautions by Stohmann, such an assumption has, it appears, but little foundation in fact. Lecturing on this subject a few years ago, Dr. F. Soxhlet stated as follows: "A clear proof that an increased supply of water is followed by an increase in the dry constituents of milk was first afforded by Stohmann's experiments upon goats at Halle. He took occasion at the same time to observe the influence on the bodily condition of the animals of an extra supply of water. A goat of 28 to 30 kilogrammes, live weight, consumed daily, from the 2nd to the 10th of May, 700 grammes of hay, 800 grammes of linseed-meal, and 3 litres of water. The quantity of milk it gave averaged 1418 grammes per day. On May 11th, the goat had, in addition to the usual solid food, 6·5 litres of water; on the 12th, 3·6; and on the 13th, 3·65 litres. The yield of milk for these three days rose from 1418 grammes to 1566, 1579, and 1618 grammes respectively. From an analysis of this milk, as well as of the quantitative determination of its constituent parts, it appeared that the increased secretion was not due to an increased proportion of water contained in it, but to an actual increase of all its constituents. The composition of the milk during the period May 11th to 14th corresponded almost exactly with that which it presented both immediately before and immediately after. In reference to this point Stohmann says: 'An increased consumption of water favours the secretion of milk, but is unfavourable to fattening; the production of milk requires a large conversion of albumen, and everything that increases such conversion counteracts fattening.' . . . From this it follows that if attempts are made to increase the secretion of milk by an additional supply of water, care must be taken to ensure also an increased consumption and digestion of food, otherwise the enlarged secretion will go on at the expense of the body generally, the animal will fall off in condition, and consequently any augmented yield will cease.

Period of Lactation

This also varies according to circumstances, the principal conditions being regularity in the hours of milking, the entire abstraction of the milk on each occasion, and, finally, the length of time which elapses between the kiddings. The generality of goats give milk in sufficient quantity to be worth drawing during eight or nine months in the year; but many will go on milking to within a month of the day they are due to kid again, provided they are thoroughly and regularly milked. Some good milkers, indeed, if they are kept from associating with the male goat, will continue in profit for two or three years. The yield may not be more than a quart a day during winter, but as the animals get on to grass in the spring the udder enlarges and the flow greatly increases, the same goats often giving their two or three quarts a day. There are probably many that would do this if their owners chose to let them; but such a system is seldom practised, and therefore common animals are not usually in profit for more than ten months in the year.

Goat's Milk with a Deposit of Blood

It is not at all unusual to hear of a case where the fresh milk after standing for a short time leaves a deposit which is streaked with blood, and people are naturally loath to make use of it in this condition, wondering what can be the cause. Now this, although alarming to the uninitiated, is really a trifling matter, the milk very often assuming its normal condition in a few days without any treatment. The trouble is generally due to the rupture of one or more small blood-vessels in the udder, but what causes this I am unable to explain. It generally happens with goats in their full flush, and a cure is almost always effected by just giving the animal for a few mornings a dose of Epsom salts in gruel, as recommended in the text for other complaints. There is nothing really unwholesome about the milk, especially if the blood-streaked sediment is left behind when used.

Milk Recording and Milk Records

Goat keepers have in recent years followed the example of dairy farmers in establishing records of the yields of their milch stock. The value of keeping records of this character are many. Where a herd of goats is kept the best milkers are distinguished at a glance, while inferior animals not worth their food can be detected and got rid of. Only the best milkers being kept for breeding, and milking qualities being transmitted to progeny, an improved strain is soon established. When stock from such strains are sold, much better prices can be secured, as figures can always be produced in proof of transmitted qualities.

Any serious reduction in the yield of one of the animals is at once noticed, and as such reduction of yield may indicate a coming or present ailment of some sort, the goat can be at once examined, and if necessary treated. In the same way, as there is usually a falling off in the milk supply when a goat comes in season, attention is at once called to this condition as a likely cause and action taken accordingly. Where an attendant is kept in charge of the herd the owner, who should frequently inspect the Record Sheets, on which entries are made, has thus a check on the attendant.

Joining a Milk Recording Society

Goat keepers wishing to record the yields of their stock must apply to the Milk Recording Society of their district to become members, and they will also do well to obtain from the Ministry of Agriculture and Fisheries their little book on "Milk Recording," which gives particulars of Regulations and Notes for their guidance.

All goats under the Ministry's Scheme have to be milked twice daily, and the result of each milking as shown on the milking balance provided for the purpose has to be entered in pounds and ounces on the Sheets. On all the sheets the writer has inspected the figures were inserted daily, but the Ministry will accept weekly weighings where it is found impracticable to take records daily. It is, however, only under the former arrangement that absolute accuracy and the full benefit of the system can be obtained.

If desired, the quality of the milk of any particular animal can be tested for the butterfat constituents and reported upon.

The cost of recording goats varies according to the Society from 3s. to 5s. per head, plus the annual subscription.

In connection with Milk Recording a very useful addition was made to the Herd Book No. 31 (January to June 1926). The "order of merit" indicated by the star,* previously mentioned, is based on milking competitions, and to be obtained the breeder must go to the expense of exhibiting. The animal itself need only show the needful yield in two milkings of one day or three milkings of a day and a half (36 hours). Now, it is well known that many a goat has gained its laurels in this way at one or two shows with heavy yields that have diminished rapidly afterwards when the forcing to obtain it was discontinued.

Recording, on the other hand, encourages a long lactation, a most important matter—and any honours to be acquired are gained at home. Of course, a large proportion of the star winners figure also in the Recording Section. This section is divided into five classes, all the goats entered therein being exceptional milkers, but in varying degrees. Thus, Class 5 (the lowest) takes goats with records of from 1,000 lb. upwards, but under 1,600 lb. during the year which begins on October 1st (or from the beginning of the lactation period of that year). Class 4 is for superior milkers with yields from 1,600 lb., but not reaching 2,240 lb.—the equivalent of a ton. No symbol to indicate exceptional merit is attached to names of animals in these lower classes. In Class 3, however, such recognition is shown by the Section mark "§." The figures here range from 2,240 lb. to just under 3,000 lb., or by measure, reckoning roughly 10 lb. to the gallon, from 224 to under 300 gallons. Class 2 includes those goats yielding from 3,000 lb. to 4,000 lb., indicated by the double "§§," while to reach Class 1 a total of 4,000 lb. must be given, such a goat qualifying for the treble "§§§."

CHAPTER XVII

GOATS' MILK: ITS VIRTUES AND UTILITIES

FROM the most remote period goats' milk has been used as food for man, as shown by the Scriptures, where the milk of the goat is mentioned more frequently than any other. At the present day, however, it is almost a scarcity, and on account of its hygienic, nourishing, and digestive properties is chiefly sought after for infants and invalids. Indeed, much has been written and published of late years, both in this country and on the Continent, on the advantages of goats' milk as a diet for young children and persons suffering from wasting diseases —consumption in particular—and many instances have been given of lives saved by its use. Still, this valuable article—although better appreciated than it was forty years ago—is a long way from holding the position it deserves in public estimation generally. A fact has, however, been brought to light in recent years—the result of scientific investigation—which still further enhances the value of goats' milk, and that to an immeasurable degree; a fact that cannot be too widely known, and the importance of which cannot be over-estimated.

IMMUNITY FROM TUBERCULOSIS GERMS

The numerous enquiries that have been conducted by scientists have elicited—what is now universally admitted—that tuberculosis can be, and is, communicated to human beings through the milk and flesh of cattle afflicted with that disease, the increase of consumption in children being largely attributed (according to Lord Playfair) to the use of tuberculous milk. Now, when we consider, on the one hand, the terrible character of this insidious and fatal disease, and, on the other hand, the absolute necessity for the use of milk in the healthful rearing of children, such a revelation is simply appalling. What makes matters worse, moreover, is that a cow may be suffering from the malady in its earlier stages without the disease being detected; for we are told that "there may be no appearance visible to the naked eye of the action of the tubercular bacillus in a particular animal, and yet it is in all probability there." Indeed, a case was reported in the *British Medical Journal* affording an illustration of this. In view of such a state of things, who will not experience a sense of relief on hearing that goats' milk is practically free from this element of danger? Such, however, is the statement of numerous medical and veterinary authorities.

The late Sir William Broadbent, in an address on the "Prevention of Consumption and other forms of Tuberculosis," delivered at the Technical College, Huddersfield, in October, 1905, made allusion to this fact in the following sentence: "It is interesting to note that asses and goats do not suffer from tuberculosis, and to bear in mind that the shrewd physicians of past days used to order asses' and goats' milk for persons threatened with consumption." Dr. Pol Demade, director of the Sanatorium at Haeltert, Belgium, in a paper read at the First National Congress for the "Improvement of the Goat," said, in comparing the milk of the goat with that of the cow in this particular: (1) There were in Belgium at one time 300,000 milch goats and 900,000 milch cows; (2) amongst those 300,000 goats there was probably not one affected with tuberculosis, whilst amongst the cows, according to whether the estimate be made by an optimist or a pessimist, there might be anything between 50 per cent. and 75 per cent. of animals suffering from or showing signs of this disease; (3) the goat is almost immune from tuberculosis.

M, Crepin took considerable pains to obtain opinions from veterinary authorities, and especially from those practising in the Mont d'Or district, where goat-keeping is general. He quotes M. Mathis, professor in the Veterinary College at Lyons; M. Coquet, a veterinary surgeon at Vaise having a large experience as inspector of butchers' meat in the slaughter-houses of Lyon-Vaise; and many others. All agree in the evidence they give as to the extremely rare instances in which tuberculosis has been discovered in the carcases of goats, for there goats' flesh is largely consumed as well as goats' milk. M. Provent, veterinary surgeon at Fontaine-sur-Saône (Rhone), is reported to have written: "I have *never* met with a case of tuberculosis amongst the goats of Mont d'Or during the whole of my twenty-four years of practice in that region." (See Appendix.)

Goats' Milk *v.* Cows' Milk

The researches of Dr. Barbellion, of Paris, on the digestibility of goats' milk as compared with cows' milk, reference to which was made at the International Congress of Medicine in that city, explain the reasons why the former is more digestible than the latter: According to the results of experiments made, it was ascertained (1) that the curd of cows' milk forms a dense, adhering mass, which, by agitation, separates into clots that are but slightly soluble. The curd of goats' milk, on the other hand, forms into very small, light flakes, which are soft, very friable, and very soluble, like those in the milk of the ass and in human milk. (2) The curd in both human milk and in that of the goat, after agitation, is precipitated very slowly and incompletely, while the curd of cows' milk is precipitated very rapidly and very completely. (3) Submitted to the action of digestive ferments, human milk and the milks of the ass and the goat were digested completely in twenty hours, whilst the same process applied to cows' milk showed only a very slight advance after sixty hours. (4) The milk of the goat approximates more in its composition and digestibility to human milk than that of any other animal.

Dr. Augustus Voelcker, the well-known analyst of the Royal Agricultural Society of England, some years ago explained the comparative digestibility of goats' milk in the course of his report upon some goats' milk he had been analysing. "The cream-globules in goats' milk," he remarked, "are smaller than in cows' milk, and as the milk is more concentrated than cows' milk, the cream-globules are contained in goats' milk in a more perfect state of emulsion than in cows' milk, in consequence of which hardly any cream rises to the surface on allowing goats' milk to stand at rest for twelve hours or longer. One of the samples threw up scarcely one per cent of cream, and the two others none at all, on standing for twenty-four hours."

In order to destroy in cows' milk any possible germs likely to generate disease in the human organism it has been the practice of late years to sterilise it, in which condition it becomes, as has been truly said, "a congealed defunct liquid." D'Escherisch, who has studied comparatively milk fresh and sterilised, has shown that milk is not merely a nutritive liquid, but is endowed with a biological activity upon the digestion, absorption, and assimilation of its component parts. "Milk," he says, "contains substances in the nature of ferments of a varying solubility according to the class of animal. These ferments, necessary to digestion, are destroyed by sterilisation. Thus it is that vigorous children with a well-formed digestive canal, and which is well furnished with digestive ferments and assimilators, are able at a push to thrive on sterilised milk, but, on the other hand, children that are delicate have need of the ferments contained in raw milk and are quite incapable of digesting milk rendered inert by sterilisation."

Virtues of Goats' Milk

Dr. William Ker Bell, M.R.C.S., L.R.C.P., writing in the Year Book of 1923 quotes the Medical Officer of Health for London, as stating that "out of fifty consecutive samples of milk purchased in the City no less than *one in four* proved to be tuberculous," and we are further told that some ten thousand children die annually from drinking tuberculous milk. Dr. Bell considers goats' milk invaluable both for invalids and people in good health, but it is for children that he regards the supply as vital. As a substitute for mother's milk we read "When cows' milk is prescribed for bottle feeding it is generally diluted either with water, lime water, or barley water, but this does not appear necessary in the case of goats' milk which is soon tolerated undiluted . . . . . Goats' milk can more safely be used unboiled (than cows' milk) on account of its greater digestibility and purity." As a corollary the following important statement concludes the article referred to: "That statistics from Germany prove infant mortality to vary in inverse ratio to the number of goats kept in different localities is very significant and demonstrates the importance of an adequate supply of goats' milk—at any rate for children." To ensure absolute immunity from tuberculosis it is important that goats should not be kept on pasture where tuberculous cows are or have been grazing.

Another medical man, Dr. Hewitt, of Montpelier, Cheltenham, wrote to me saying: "I fully agree as to the superior value of the milk of the goat, and can write positively as to the advantage of the same for the nurture of the human infant, since all my children were brought up by hand, and for the first six months of their life their staple food was the product of the goat. A marked change was speedily noticed in their aspect when occasionally, from necessity, cows' milk had to be substituted. Moreover, I know from experience that goats' milk is, in very many cases, more easily assimilated and has a better nutritious value than that of the cow in all low conditions of vital force and debility from chronic wasting diseases."

Goats' Milk for Domestic Use

Apart from its medicinal qualities, however, goats' milk is, for domestic purposes alone, far superior to the ordinary milk supplied by dairymen, as all who have tried it can testify. Boiled and used with coffee it is delicious, giving the latter a rich creamy appearance, whilst a few drops in a cup of tea are more than equivalent to a teaspoonful of cows' milk. When used in cakes and puddings, its superiority is quickly apparent to the palate, imparting a rich flavour to these articles when cooked, which can always be recognised by those who have once tasted it. The only disadvantage of goats' milk for cooking purposes is its liability to curdle, which it is very apt to do if used rather old. I have seen this statement disputed somewhere, but such at least is the experience of those in my household.

The useful qualities of goats' milk only require to be better known to be properly appreciated. As it is, for want of being known and understood there has existed a general—I might almost say universal—prejudice against this valuable article of diet, people supposing that the milk has a peculiar flavour, which is quite a mistake, for when drawn *clean* from an anima in health it resembles cows' milk, both in taste and appearance, the only difference being that it is richer, thicker, and slightly sweeter, containing as it does a larger proportion of cream (butter-fat), and sometimes more sugar, as will be seen from the analyses given below.

The quality of the milk varies with the season and according to the length of time from kidding, being less concentrated when the goat has kidded a month or so than when she is going dry. At each milking, even, it is not of the same consistence throughout, those portions which are first drawn from the udder containing less butter-fat than the last drops, commonly called the "strippings," which are nearly all cream. If these be put to stand in separate vessels the difference in quality will be readily noticed.

Chemical Composition of Goat's Milk

This varies very considerably in different individuals as also in different breeds. The following table shows a fair average sample of the milk from a cross-bred goat, winner of a second milking prize at the Dairy Show, analysed and compared with other milks of average quality. It shows this sample of goats' milk to be richer in butter-fat than all the other milks and only slightly below that of the ewe in general solids :

	Goat	Ewe	Cow	Ass	Human
Casein	4.06	5.37	4.48	1.82	1.52
Butter-fat	5.14	3.65	3.13	1.10	3.55
Sugar	5.28	5.46	4.77	6.08	6.50
Salts	0.58	0.79	0.60	0.34	0.45
Total Solid Matter	15.06	15.27	12.98	9.34	12.02
Water	84.94	84.73	87.02	90.66	87.98
	100.00	100.00	100.00	100.00	100.00

The Nubian goat, like the Mamber or Syrian, has the character of giving milk very rich in butter-fat, and this quality is shared by the Anglo-Nubian as a rule, whilst the Toggenburg and other Swiss breeds yield a milk that is always poor in this constituent. A good illustration of this fact, which has been proved by repeated experiments, is shown in the following table, extracted from the analyst's tabular report of the milks of goats competing for milking prizes at the Dairy Show :

Name of Goat and Breed	Sedgemere Capella Toggenburg	Sedgemere Faith Alpine	Sedgemere Louise Anglo Nubian
Days in Milk	183	189	635
Yield of Milk in lb.	9.1	7.9	3.3
Percentage Composition of the Milk :			
Fat	3.21	3.64	7.48
Solids other than Fat	8.05	8.04	9.46
Total Solids	11.26	11.68	15.94
Water	88.74	88.32	8.306
	100.00	100.00	100.00

The total solids in the milk of the Anglo-Nubian in this table, however, are extraordinary, even for goats' milk, and might be explained by the fact that the goat had been a very long time in profit ; but the same animal the previous year, when it had been in milk only 179 days, gave the following figures : fat, 7·52 ; other solids, 9·32 ; total solids, 16·84 ; leaving only 83·16 for water. At this show the same goat won first prize, besides the Baroness Burdett-Coutts's Challenge Cup as the best for inspection and milking combined. Another Anglo-Nubian—to take a more average specimen—named Sparaxis, which was placed after the third-prize winner in this competition, gave milk showing the following figures : Butter-fat, 5·81 ; solids other than fat, 9·47 ; total solids, 15·28 ; water, 84·72.

In M. Crepin's book "La Chèvre" a number of analyses are given of the milk of various breeds of goats under specified systems of feeding. In most of the cases examined the milk of French or Swiss Alpines and of other breeds of Swiss goats shows the butter-fat to be under 4 per cent. It may here be stated that 4 per cent. has been fixed by the British Goat Society as the minimum allowable in awarding points for quality without deductions, in the same way that 3 per cent. is the Dairy Show minimum for cows' milk. The poorest milk in butter-fat in the tables just referred to is that given by a herd of Swiss Alpines, for which only 2½ per cent. of this constituent is recorded. In another herd of this breed, containing fifteen head, all of which were it is stated, subjected to a special regimen as regards food in order to obtain a milk the composition of which approached more that of human milk, the butter-fat is given as 4 per cent., the casein as 1·73 per cent., and sugar 5·4 per cent. The richest in butter-fat of all the breeds examined is that of the pure Nubian, three samples of different individuals showing respectively 5·8, 4·6, and 8·4. This last, however, is quoted from Huart du Plessis' book.

The richest sample of goats' milk I have personal knowledge of came from an Angora which was once in my possession. This milk was analysed by Mr. F. J. Lloyd, late analyst to the British Dairy Farmers' Association, and gave the following extraordinary figures: Butter-fat, 8.69; casein and milk-sugar (being the "other solids"), 9.85; mineral matter (ash), ·93; and water, only 80·53.

General Remarks

Milk, containing as it does fatty particles in a state of suspension, readily absorbs all kinds of odours, and should therefore be kept apart from everything that can influence it in this way. For this reason it should be removed from the goat-house as soon as possible after being drawn, and put in a cool, airy place.

The milk that is collected in the udder at the time of parturition is of a different character and consistence from that which flows a few days after and subsequently. It is then highly coloured and thick, owing to its containing nearly three times more casein than at a later period; it has only a trace of sugar, no salts, but a considerable proportion of mucus. In this condition it is known as "beestings," and is not really fit for ordinary domestic use, though some people make a kind of custard with it by placing it directly after being drawn in a pie-dish in the oven, when it readily coagulates, and on being cooked has certainly a resemblance to a custard in appearance, though not in taste. Although not a fanciful dish to grown-up persons, children generally like it, especially when eaten with some preserve. It is by no means unwholesome, but to the newly-born kid it has a slight purgative effect.

All milk when first drawn, and so long as it contains its natural warmth, is said to possess a peculiar flavour, varying according to the animal by which it is produced. If shut up in this state in a closed vessel, the animal flavour is to a great extent permanently retained, and the keeping qualities of the milk are impaired. It should therefore in all cases, but particularly when sent by post or rail, be gradually cooled down before being placed into the closed receptacle intended to convey it. When a large quantity is being handled it is best to use a refrigerator for this purpose; but if a small amount only has to be treated, it may be aerated merely by being poured from one vessel to another in the open air, and then allowed to stand in a shallow pan surrounded by water until quite cold.

As regards its keeping properties, goats' milk is not equal to cows' milk, as it more easily curdles on boiling if not quite fresh. Although if new and good it is free from any flavour or smell of a hircine character, if some drops are spilt, especially on wood, and not wiped off, in a few days, when decomposition has taken place, a strong goaty scent will be found to emanate from the part where the milk was spilt. In the same way a very distinct odour of a similar kind may be noticed in some soft cheeses made from the milk which, in the ripening of the cheese, has undergone decomposition.

If goats' milk is boiled whilst still quite fresh, it will keep longer than when in a natural state; but the flavour and some other properties are detrimentally affected by the action of heat. Its keeping qualities may be considerably extended by stirring in about two grains to the pint of salicylic acid, which has the advantage of being both tasteless and inodorous.

It is a singular fact that goats' milk is not usually influenced in its flavour by the food the animal has consumed in the same way as cows' milk. Thus a goat may feed on such strong-tasting articles as turnips, laurel, ivy, and such like without imparting a corresponding flavour to the milk.

Unwholesome Milk

At times the milk that comes from a goat will be found tinged with red, especially towards the strippings. When such milk is allowed to stand for a short time and is then poured off, there is a distinct sediment of matter and blood. This condition may be owing to a chronic and localised inflammatory action going on in some part of the udder, which has resulted in suppuration, and the discharge of matter and other inflammatory products into one of the milk-ducts instead of the udder. Although cases of this kind are sometimes sufficiently serious to require veterinary aid, I have frequently had instances when the milk has assumed its normal condition in the course of a few days without any treatment beyond giving a dose of Epsom salts to the goat each morning.

Goats' milk, like other milk, may be poisoned by the animal eating some poisonous plant without being itself affected. Although this is very rare, and has never happened in my experience, cases were recorded in the *British Medical Journal* some years ago as having occurred in Rome. The goats in question had been pasturing in the neighbourhood of Borgo Rione, where four poisonous herbs were found on investigation to abound, viz., *Conium maculatum*, *Clematis Vitalba*, *Colchicum autumnale*, and *Plumbago europæa*. The persons who had partaken of the milk from these goats suffered from a kind of cholera, and although recovery took place in each case, the patients were seriously ill for several days, the severity of the symptoms being in direct relation to the quantity of milk taken. The goats themselves, strange to relate, remained perfectly healthy. On analyses of the milk of the animals and the vomit of the persons attacked colchicum was discovered in both, its passage into the milk being regarded as the cause of the poisoning.

Goats' milk in connection with the "Malta," or, more correctly, Mediterranean, fever will be found dealt with in Chapter XXVI.

Casein-like Formation in Milk

Sometimes the udder of a goat will produce, each time she is milked, a white curd-like formation, which chokes up the orifice of the teat and causes much trouble to the milker and some pain to the goat. A sample of this substance was submitted by me to the analyst of the British Dairy Farmers' Association, Mr. F. J. Lloyd, who wrote in reply: "The material which you sent me as coming from the goat is most interesting. When I first saw it it appeared to me to be a little caseous matter, but I was surprised when I examined it under the microscope to find that it was not casein, but consisted of a mass of cells such as I have seldom, if ever, seen before. The cells are a very great size and very regular in shape, but most devoid of internal structure. The nearest approach to them that I know of are the epithelial or squamous cells found in the mouth, but these are nearly always irregular in shape, while the cells of the material you sent me were most regular. After examining the material, I sent it on to Mr. Cecil Revis, thinking it might interest him, as he is engaged in an investigation of the cellular contents of milk. He writes: 'It is the kind of cell which I have noted once or twice in the milk of certain cows. I say the word "cell" in a careful manner, because I cannot settle at all what these bodies really are, and I have shown them to one or two other people who have also not been able to give them a name.' Personally, I think it highly probable that these cells may have lost their contents, but the cell-wall was not broken down, as is generally supposed to happen in milk-formation. If the trouble continues I should like to be put in direct communication with the owner of the goat, as I think the subject is worthy of a little investigation—it is certainly rather exceptional."

Goats' Milk for Puppies

Goats' milk seems particularly well adapted for feeding puppies. Although in making this statement I do not speak from my own experience, I have nevertheless good authority for the assertion, it being founded on the results of trials made by several prominent dog-breeders, some of whom believe that puppies fed with this milk are much less liable to be troubled with worms—that pest of all breeding-kennels—than those fed with cows' milk. A correspondent, writing to the *Live Stock Journal* some years ago, stated that after losing through these parasites 75 per cent. of the puppies he bred, the cause of which he attributed to the use of cows' milk, he was induced to try goats' milk, which proved so successful that, at the time he wrote, he had bred more than fifty puppies without one showing a sign of worms.

Considering that there is a wide difference in the constituents of goats' milk compared with the milk of the bitch, which contains not only more casein, but also more sugar, it is somewhat remarkable that the former should agree so well with puppies. In an article devoted to this subject in the kennel section of the journal just referred to the writer stated: "In looking over the table of the different constituents of the two milks, in juxtaposition to the milk of the bitch, it is evident at a glance that although the differences are still great, yet the composition of the goats' milk more nearly approaches the one of the bitch except in the quantity of casein; and for that reason renders goats' milk a more desirable food for puppies, being less susceptible of disturbing the digestive organs, and easier of assimilation. The greater proportion of fat, taken in conjunction with the lesser proportion of casein and sugar, undoubtedly renders goats' milk less indigestible in minimising the quantity of curd formed in the stomach. . . . Another reason suggests itself why goats' milk should be a natural food of better quality than cows' milk, and it is this: The proportion of solids is greater in goats' milk than in cows' milk, and as those solids are in part constituted by soluble salts, such as phosphates of lime, of magnesia, and of iron, with chlorates of potassium and of sodium, and as these soluble salts are required for the formation of bone, the feeding of puppies with goats' milk gives the little animals the required substance in a form similar to the one they were having from their mother's milk."

Mr. W. K. Taunton, who made his name as a mastiff-breeder from the remarkable litters of puppies he used to exhibit and which carried off so many honours, informed me that he attributed his success in this respect to the use of goats' milk in his kennels, having given it to his young dogs whenever he could obtain it. As Mr. Taunton kept goats for this purpose only, it is sufficient proof that he believed in the utility of the milk. I should add, however, that this breeder did not endorse the statement made above, that puppies fed on goats' milk are free from worms, having found that most young dogs have them more or less, no matter how fed.

Goats Sucking their Own Milk

A goat will sometimes contract this objectionable habit through her milking being neglected for a long time, and also, I believe, if allowed to become very thirsty. Whatever cause may induce it, the habit is sometimes difficult to eradicate. The practice with some people is to smear the teats with ointment in which some strong and nasty-tasting stuff has been incorporated, such as creosote or asafœtida. Care, however, must be taken not to use any chemical that would act on the surface of the teat and cause pain or injury, and of course it will be no good to attempt to use the milk as long as the ointment remains on, because in the process of milking, unless the teat be washed each time, and no attempt at stripping made, the milk will naturally acquire the nauseous taste intended to disgust the goat.

Another and better plan consists in putting on the goat a loose collar composed of short sticks sharpened to a point at each end, and called a "cradle." These have a hole at about 2in. from the points, through which a thick string is passed and fastened firmly to each stick to prevent it from sliding, the latter being about 1½in. distant one from the other. When placed on the goat the sticks lie parallel with the animal's neck, but as she turns to reach her teats her neck comes in contact with the sharp ends and prevents her from attaining her object. Collars like this are also used on young bucks to cure them of objectionable habits.

A third arrangement, and this is as effective as any, consists in fixing on the goat a head-stall the nose-band of which has sharp spikes projecting from the front. These spikes are no inconvenience to the animal whilst grazing, but as soon as she attempts to reach her udder the points compel her to desist.

Secretion of Milk without Previous Parturition

Instances frequently occur, though they are not often published in which a goat gives milk without ever having borne kids or even having been mated, and such a flow may be maintained for a considerable period. The cause of this cannot well be explained, but it must be due to excitation of the mammary gland in some way or other. Thus, if the goat contracts the habit of sucking its own teats, as stated above, a flow of milk may soon be artificially induced; and, similarly, if the teats of one of these animals are frequently handled, as in the action of milking, the same result will probably accrue. When such cases occur it is generally with quite young goats, but a letter received recently from a lady residing in Aberdeenshire tells of one in connection with an aged goat. My correspondent writes as follows:

"A friend of mine was given a handsome she-kid fourteen years ago to draw a child's goat-carriage, and since then she has lived contentedly in the field with the cows. The animal has never had any kids, no he-goat having been in the neighbourhood; but last year her udder was observed to be very much inflated, and on being milked she yielded a quart and a half. She was milked once a day for a short time, and then it was discontinued, as no large quantity came after the first milking. This summer, however, the teats became again much inflated, and over *two quarts* were obtained at the first milking. The milk has no smell, and looks perfectly good, but no one has ventured to taste it. The goat herself appears to be in perfect health. Is this at all a usual occurrence?"

With a goat at the late stage of life stated this is certainly a most unusual occurrence; but in these cases there seems no valid reason why the milk should not be utilised in the ordinary way for domestic consumption, after a regular flow has been encouraged, though the first accumulation in the udder would doubtless be better thrown away.

CHAPTER XVIII

GOATS' BUTTER AND CHEESE

BUTTER

May be made from goats' milk in precisely the same way as from cow's milk, but it has not that superiority that the milk has to cows' milk; nor is it so nice-looking, being perfectly colourless. As, moreover, it does not keep quite so well, there is nothing gained in my opinion by churning it. Where several goats are kept, however, and all happen to be in profit at the same time, more milk will probably be supplied than can be utilised in its natural state, and it may then be advantageous to make it into butter.

M. Crepin speaks very highly of goats' butter, and uses no other. He tells us in his book that a pound of goats' butter has one-fifth more bulk than the same weight of cows' butter, and therefore goes much further. The same author mentions that the Comtesse de la Boullaye makes large quantities of this article, which she sells on the coast of Brittany at double the price realised for ordinary butter. I have myself tasted various samples of goats' butter, and frequently had it made in the working dairies at agricultural shows; but I can only remember once finding it to my liking. The fact is it requires to be made with the greatest care, more especially in the elimination of all the butter-milk, for if even a small quantity be left in the pats when finished they rapidly acquire a peculiar "goaty" flavour, which, to most people, is extremely unpalatable.

Those who desire to experiment in a small way with butter-making may do so by carrying out the following somewhat primitive method. It was practised regularly by a lady friend of mine years ago, who, with the milk of two goats, used to make every week during the summer when the animals were in full profit about 3lb. of butter, and who never cared to relinquish the bottle system for more elaborate apparatus :

The milkings of each morning and evening, after being strained through a fine hair-sieve or butter cloth, are set in small pans about 4in. in depth, holding each about 3 pints, and left for twenty-four hours for the cream to rise, being placed in a dairy or other cool spot free from any kind of smell. Every morning and evening after this process has been gone through, the milk which has stood for twenty-four hours is placed on the kitchen range or stove (a closed stove being preferable) over the boiler, or as far as possible from the fire, and there allowed to remain until the surface becomes slightly wrinkled, when the pan is removed to the dairy. And here it should be observed that the cooler the range or stove and the longer the milk takes to scald, the thicker will be the cream.

The next day, the cream, which has in the meantime risen, and which should be a quarter of an inch thick, is skimmed off and placed in a white earthenware jar or basin. Twice every week this jar or basin is stood in hot (not boiling) water until the cream is lukewarm, when it is poured into a wide-mouthed glass bottle, holding about a quart, a little annatto or other butter-colouring being added, drop by drop, to obtain the required tint, without which the butter will be as white as lard. This bottle, which answers the purpose of a churn, is securely corked, and shaken backwards and forwards in a slanting position from right to left (this movement being less fatiguing than any other) for twenty minutes, by which time the butter is churned. It is then, after the sour milk has been strained off, turned into a cloth, the ends of which are gathered together as in tying up a pudding, and put into cold spring water—running water for preference—where it is constantly pressed with the hands until all the butter-milk is extracted. This is important, for if any be left in, the butter will soon become rancid.

The butter is now turned on to a board, or, better, a marble slab, where it is spread out an inch thick, sprinkled with salt, and worked about well for a minute or two. Finally, a print or box mould, after being dipped in cold water, is used to make the butter up into little pats of about 2oz. each. If a box mould is used, the butter is slightly spread over the print-piece first, then drawn down over the side and spread round the middle part being filled last. The mould is afterwards placed on a dish, the print pressed firmly, the box gradually drawn up, and the print quickly removed. The pats are thus evenly formed, though a little practice is necessary to acquire the knack of making them without holes.

In summer the milk should stand only twelve instead of twenty-four hours before and after scalding. The effect of this latter process is to separate the whole of the cream from the milk, and greatly to facilitate its conversion into butter. Indeed, without this treatment the milk does not keep nearly so well, and the cream takes double the time to rise.

Besides being made from scalded cream in the way I have shown, butter may be manufactured from goats' milk in a whole state. This has been done at various times at the Dairy Shows at the Agricultural Hall with the milk from the goats exhibited there, and a very fair article produced, though not in my opinion equal to that made from the milk of the cows. Butter obtained from whole milk is more difficult to make, and not so good as when made from scalded cream. In churning the whole milk slightly more butter results, in consequence of a certain proportion of cream being retained in the skim milk after the bulk of the cream has risen ; at the same time there is a proportion of casein or cheesy matter, which, being a nitrogenous substance, is subject to rapid decomposition, thus injuring the keeping qualities of the butter.

As a rule, 12qts. of good goats' milk will make 2½lb. of butter, so that a goat yielding 2qts. daily will supply a pint of milk a day, besides about 2lb. of butter a week.

Cheese

Wherever goats are kept in considerable numbers, as is the case in Norway and Sweden, in some parts of France, and other countries bordering on the Alps and Pyrenees, the manufacture of cheese from their milk, either alone or mixed with other milk, is extensively practised.

The varieties of goats' milk cheese are numerous, and differ considerably in taste, some being extremely unpalatable owing to the admixture of foreign substances—treacle, for instance, which forms a prominent ingredient in the manufacture of one kind of goats' cheese in Norway. It is not, however, with such that I propose to deal here, but to describe those which are made and regarded as delicacies abroad, and which would no doubt command a ready sale in this country if properly made and put on the market. Such are Roquefort, Mont d'Or, and two less known in England, though highly appreciated in France under the names of Levroux and St. Marcelin. Some readers may wonder why Gruyère is not included in this category, and will be surprised to hear that Gruyère is not made from goats' milk, as so many people seem to imagine, but from cow's milk.

ROQUEFORT CHEESE.—Roquefort is made with a mixture of goats' and ewes' milk. Its manufacture was originally restricted to the plateau of Larsac, where several thousand sheep and goats were kept for the purpose; but the industry has since been extended over a much greater area.

The milk, after being strained through a cloth, is poured into a copper lined with tin, where it undergoes a certain heating to prevent it from turning sour. Rennet is then added to separate the curd, the process being assisted by agitation of the mass, and the whey withdrawn. The curds are then cut up with a wooden knife, worked up into a sort of paste, and placed in cylindrical moulds of earthenware, with holes at the bottom. After the moulds have been about one-third filled a thin layer of mouldy bread is introduced, which is followed by another quantity of curd, upon which more bread is sprinkled, and so on to the top of the mould. This *pain moisi*, as it is called, is added in order to produce the blue veins which are so highly esteemed in this cheese, and which are due to the propagation of a kind of fungus (*Pencillium glaucum*) out of the bread, which is specially prepared for the purpose.

The moulds being filled, the cheese is closely pressed in them, fresh curds being taken to fill up any space rendered vacant by the compression. The cheeses are then put into wooden boxes and left in a warm, damp place, where they are turned twice a day. From here they are taken to the drying-room, where they are wrapped round with linen bands to keep them from falling to pieces, laid on boards, and left to dry, being frequently turned. The required desiccation is effected in from fifteen to twenty days, after which they undergo a salting process. This consists in sprinkling salt upon one side of a cheese, and placing other cheeses upon it, to the number of five, and the next day reversing them and repeating the process on the other sides. They are then, after a couple of days, rubbed over with a rough cloth to work in the salt, and placed in caves, still in piles of five, where a kind of glutinous mould forms upon them termed rebarbe, a substance that has some value, and is sold as a tonic and stimulant to the stomach. This is scraped off, and the cheeses are replaced in heaps as before for a fortnight, when they are separated and ranged in rows on their sides, and apart from each other. In this position another coating of a different kind forms upon them, which, after being scraped off, is soon followed by a third, white and blue in colour, and subsequently a fourth, this time white and red, both being removed as before. The manufacture of the Roquefort is at this stage regarded as complete, though a scraping process takes place every two or three weeks whilst the cheeses are in the caves, and until they are sold.

These caves play an essential part in the successful manufacture of this cheese. They are for the most part naturally-formed grottoes in the calcareous rocks or mountains which surround the village of Roquefort, and are full of fissures, through which constant and strong currents of air blow, producing an icy coolness in the place—an important feature. The various rooms in which the different processes are carried on are built over these caves.

It should be stated that at the present day the mixture of goats' milk in the manufacture of Roquefort cheese is less practised than formerly, partly, it is said, on account of the comparative scarcity of goats' milk for the purpose, and partly because it is found that cheese made solely from ewes' milk keeps better. Something like 650,000 sheep are pastured in the neighbourhood of Roquefort for this industry, and, having been bred for years specially for the purpose, their milking qualities have greatly improved, the yield from some of these animals being nearly equal to that of a good goat.

Mont d'Or Cheese.—Compared with the cheese mentioned above, Mont d'Or is but little known in this country, though a great favourite in France, particularly about Paris and Lyons.

These cheeses are thus made: The morning's milk is placed in earthenware pans, and the rennet added and well stirred in with a wooden spoon. When the curds have formed, which occupies about a quarter of an hour in summer, and double that time in winter, they are poured on clean cloths stretched over round boxes, placed on straw, and left to drain. On being removed, the cheeses are again laid upon straw, and salt is sprinkled over them. The next day they are turned, the cloth is replaced by a clean one, and the other side of the cheeses salted. This process is repeated for several days until the cheeses become sufficiently firm. They are then transferred to wicker baskets and suspended to the ceiling by means of a cord passed over a pulley. Here they remain for about fifteen days to dry, after which they are fit for consumption. It is a common practice to soak them in white wine for a day or so before being eaten; and, in order to soften them, they are turned over daily, so as to lie on each side alternately.

Levroux Cheese.—At the little town of Levroux, in the department of Indre, some excellent cheeses are made which deserve to be better known. They have the form of small square bricks, and are composed of three parts goats' to one part cows' milk. They are made very much like those last described, but before being exposed for sale they are treated in a very singular manner: that is, they are steeped several times in melted butter in which onions have been fried, the butter being then passed through a woollen cloth to remove all particles of onion. They undergo this treatment of soaking two or three times in the course of a week, the object being to soften them and improve their flavour. The cheeses are subsequently enveloped in chestnut-leaves or vine leaves, and eight or ten of them placed together in a brown earthenware jar.

St. Marcelin Cheese.—The late Sir Howard Spicer, while travelling in the South-west of France, had occasion to stop in the Department of Isère, whence he sent me an account of the cheese-making process from which I extract the following:

"The daughter joins us with three large jars for milking, and they think it a fine joke when Monsieur offers to milk 'Blanchette,' a large Alpine and a milker all over. Monsieur gives her a piece of bread, rubs her cheek and neck, scratches her head, and as she is munching makes a start, and as she nibbles his ear the steady flow into the big cool jar makes grand'mère exclaim, 'Nom d'une pipe, mais monsieur aime bien les bêtes et elles le savent !'

"I have paid my footing, and the mysteries are at my command. Back we troop laden, and the news that Monsieur has milked 'Blanchette' is passed around ; the milk is heated, put back in fresh jars, and the rennet added, a teaspoonful to the litre.

"Then the morning's brew is taken, and with a flat ladle, full of holes, the curds are put into round tin pots, full of holes also, about 4in. deep and 4in. wide, and these are placed on a huge sort of soup plate for draining. Those that have been draining for twelve hours are salted top and bottom with coarse salt and put back to drain further ; and those that are salted and twenty-four hours old are put into long flat baskets full of straw, and hung on the roof of the larder to dry, away from the flies and insects.

"And that is how they make 'fromage St. Marcelin.' Eat it new or matured, as your appetite prompts you."

I may here remark that I have more than once tasted English-made St. Marcelin cheeses, and have found them excellent, but perhaps the best proof that they are good and have no goaty flavour consists in the fact that this dairy produce was in great demand in the kitchen at Sir Howard's house.

RUFFEC CHEESE.—This is another cheese made in France, for the recipe for which I am indebted to the same source. It takes its name from the town which is the centre of the district in which it is made, about thirty miles north of Angoulême, in the Charente. This variety is made from the milk when it is cold, and the quantity of rennet required is somewhat less than that needed for the St. Marcelin. After the curd has set it is put in a muslin bag and hung up for forty-eight hours, being turned over so that the weight comes on the other side every twelve hours. After it has been hanging up for about twenty-four hours a little salt is sprinkled over it, then it is turned again. It keeps good for about eight days after making, and is eaten like a cream cheese.

CHAPTER XIX

THE FLESH OF THE GOAT

ALTHOUGH the flesh of goats generally—and especially that of the Angora—is largely consumed and preferred to mutton in countries where these animals abound, I should never recommend keeping ordinary kinds for this purpose; for so long as we have beef and mutton in plenty few people will care to eat goat, besides which our present breeds are too slow-growing to be adapted for the purpose. There is, however, more prejudice against this meat than is really warranted, for many people might eat of a goat eighteen months or two years old and never realise that they were not partaking of mountain mutton. I qualify the remark in this way because, as a matter of fact, I am told a great many carcases of goats find their way into the metropolitan meat markets, to be there sold as "house lamb," though fetching of course an inferior price to the genuine article. Having been myself very nearly victimised in this way, I can speak from experience. The circumstances of the case being somewhat amusing, besides serving as a caution, I may as well relate them here. It happened at a time of year when lamb fetched a high price, and when the temptation to offer a spurious article was therefore very great.

Passing a butcher's shop where some joints were exposed for sale marked at a comparatively low figure, I entered and ordered a hind-quarter. The serving-man took a joint from the interior without my taking particular notice of it, weighed it, and packed it in a rush-basket, and I carried it innocently away. Arrived at home with my "lamb," I was called into the kitchen and asked in a somewhat quizzing manner if I was aware that I had brought home some goat for lamb. Such an assertion I of course at once repudiated, but on examination of the hoof that hung to the leg, and which was covered with brown hair without the slightest resemblance to wool at any part, together with the dark colour of the meat, the absence of fat, and diminutive size of the chops, the matter was soon placed beyond doubt.

I accordingly returned next day with the meat to the butcher, and gave him very plainly to understand that he mistook his customer in attempting to palm off goat upon me for lamb. The man of course would not admit the fact, and indignantly denied the charge, endeavouring to show me that the hoof had no resemblance whatever to that of a goat. With the view of proving his statement he called out several times for "Bill." I naturally expected to see the summons answered by one of his men, but in a few minutes, to my surprise, in trotted a medium-sized billy goat, and I was requested to compare his hoofs with the one on the leg I had bought and to observe the difference. I made the comparison, but only to be confirmed in my opinion by the close resemblance they bore. In fact, I could not help looking to see if the animal under examination had a leg missing. This similarity Mr. Butcher refused to acknowledge, remarking that I had evidently *never seen a goat before!* The fact, however, of a goat being thus on the premises in this part of London only served to strengthen my suspicions.

An assistant, who was busy with the meat outside, and had apparently taken no notice of what had passed, was then called in by his master and thus addressed, "Now then, Tom, *you* know what lamb is when you see it; what do you call that—lamb or goat?" The man, under the impression, I suppose, that I had brought the meat into the shop for a professional opinion, after turning it over several times, said in a very decided manner, "Why, goat, of course."

I was much amused by the annoyance and disappointment depicted on the master butcher's countenance, and the way he invited his man to re-examine it in order to elicit a different opinion. This, of course, was soon obtained when the latter saw what kind of answer was expected of him. I told them, however, I considered the question settled, and that I must insist on my money being returned, as it was evident that the joint was not lamb. This was presently done, the butcher consoling himself by repeating the remark that I knew nothing of either goat or lamb, a point I did not deem it worth while to dispute.

The meat of a full-grown goat has more resemblance in appearance to venison than to mutton, being closer in the grain and also of a darker colour than the latter. As I have elsewhere remarked, the fat of these animals accumulates almost entirely at the kidneys, a very small proportion being deposited in the subcutaneous cellular tissue; the consequence is that this meat is somewhat dry eating, and requires well basting.

Castration in the case of the male has a very marked effect upon the quality of the meat, producing a finer flavour, and greatly increasing the quantity of fat. I once had killed an emasculated buck two years old, weighing as dead meat and "sinking the offal" 40lb., and it was highly appreciated by all who tasted it, some mistaking it for venison. If is of course better that castration should be done while the animal is young, but it is not uncommon in some countries, and especially in Algeria, to perform this operation after the goat has had three or four years' service, and then to eat him.

The flesh of the Angora is considered far superior to that of other breeds, and quite equal to mutton. Some experiments were made by M. Chevet at the instance of the Zoological Society of Paris some years ago which are well worth recording here. The subject of the trial was a male Angora which had been used for stock-getting for two years, and was emasculated at the age of three and a half, so that circumstances were somewhat against a favourable verdict, and were not improved by the fact that the goat had been imperfectly fattened. The meat, however, presented a very good appearance in its raw state, and when prepared in various ways gave perfect satisfaction, being considered quite equal to mutton, though not, of course, of the first quality.

"This meat," writes M. Chevet, "is distinguished by a decidedly sweet taste, especially noticeable in the kidneys, which were grilled, and in the broth made from the neck and breast. The leg was cooked like a quarter of venison—that is to say, larded and pickled for three days—and then roasted an hour and three-quarters before a moderate fire. The saddle was braised, and though the cooking only took two and a half hours the meat was overdone. The shoulder served as cold *rôti* was prepared *en daube farcie à la gelée*: this jelly in particular left nothing to be desired for fineness of flavour, although assisted by no seasoning whatever. Every part of the animal was tried without special preparation, and fourteen of my colleagues were convinced with me that the flesh of the Angora is perfectly good eating, and under favourable conditions should be equal to the finest mutton."

Cured Goat's Meat

It was at one time the practice in Wales to cure the hind-quarters of a goat in something after the same fashion as hams, the meat under this process being called *coch yr wden*.

I have never attempted the experiment, but a lady once gave me a recipe for *jambon de chèvre* which she "tried with success upon a goat past its youth, and that had been used in harness." This lady wrote: "The result was such that I believe with better meat it would be superfine, and supply a real relish to the breakfast-table as long as any remained. When a goat of your herd is about to be killed, have ready in a large earthenware pan the following liquid:—Water, 3 gallons; pearlash, 1 scruple; common salt, 5lb.; bay salt, 3oz.; saltpetre, 1½oz.; mixed spice, or any one favourite ground spice, 1½ teaspoonfuls. The bay salt and saltpetre to be pounded and all to be boiled together, skimmed, and set to get cold. Lay in the salting-solution as many joints of goats' flesh as it will cover. Turn them over about once or twice in the course of three weeks, and take them out at the end. Give your friends a slice of this meat when boiled, and they will ask for another. It is especially good cold. If you wish to salt the whole goat, as will be necessary with an old one, start with double the above quantity in two pans, according to size of animal. But even of an old goat the liver makes an excellent dish of fry; and of a moderately old one the head cooked with vegetables after any recipe for sheep's head would please a Scotch taste."

Kid Meat

Whatever may be urged against the flesh of the goat, nothing but the strongest prejudice can account for kid being so little eaten in this country. In Italy, Spain, and the South of France it is in constant demand, and all who have lived in Malta and the East Indies can testify to the appreciation of this meat in those countries. With the Hebrews and Greeks it ranked amongst their most dainty dishes, and everyone knows how frequently it is spoken of in the Old Testament.

To be enjoyed in perfection a kid should be killed before it has been weaned, or, more properly, before it has commenced grazing; in other words, when about three weeks or a month old. The flesh is then very white and delicate, and when eaten cold with ham would never be distinguished from chicken. At this age the meat can hardly be said to resemble lamb, not being so close in the grain as the latter (lambs are never, of course, killed at this early period of their lives). At three or four months old the flesh of a kid might be taken for lamb or veal, great though the difference is between these meats. A roast leg, for instance, served with mint sauce, might very readily be mistaken for the former, whilst the loin with rashers of bacon tied round and stuffed as for roast veal would lead most people to suppose they were partaking of the flesh of the calf. I have several times practised these little deceptions at my own table on persons who I knew would be prejudiced against the dishes if presented to them without this disguise, and always with complete success. Most people prefer the "veal" imitation to the "lamb," as the meat, not being very full-flavoured and somewhat devoid of fat, is more tasty with the addition of stuffing and with the slices of fat bacon cooked with it. In any case it should be well basted. The remains of the joint served as a fricassée the next day make a most delicious dish.

A PUBLIC KID DINNER.—In the year 1880, with the view of bringing the flesh of the kid into more general use, and, at the same time, of practically illustrating the various methods of cooking this animal, a public dinner was organised by the Committee of the British Goat Society at the Alexandra Palace on the last day of the Goat Show held there in July. Not only kids, however, but full-grown goats were killed for the purpose, and cooked to represent certain well-known dishes which are usually made up of other meats. Mr. Batchelor, the chef who had the management of this dinner, undertook to utilise every portion of the animal from the head down to the hoofs, and this was carried out to the letter, neither the flesh nor produce of any other creature being admitted, beginning with the soups and ending with the ices. Thus the meat, hoofs, suet, milk, cream, and butter of the goat or kid were each and all utilised in the various courses, the menu being as follows :—

POTAGES

Scotch Broth. Billygoat-tawny.

De Galle.

ENTREES

Kari de Lapereaux à la Madras.

Vol au Vent de Quenelles de Volaille.

Fricandeau aux Tomates Farcies.

Cotelettes d'Agneau aux Concombres.

RELEVES

Boiled Legs of Lamb and Spinach.

Mutton and Mushroom Pie.

Roast Fore-quarter of Lamb and Mint Sauce.

Steak and Oyster Puddings.

Chicken and Truffle Pies. Lambs' Heads Maitre d'Hotel.

ROTI

Roast Hares bardées. Red Currant Jelly Sauce.

ENTREMETS.

Calves' Feet Jelly. Custard.

Vanille Creams. Hot Lemon Pudding.

Ice Pudding.

It will thus be seen that amongst these various courses kids' or goats' flesh was served to imitate beef, veal, lamb, mutton, chicken, hare and rabbit, and the resemblance in some cases was so striking that only an epicure could have detected the difference. I have thought it worth while to give the menu in full, because of the great success achieved and the universal notoriety the affair obtained at the time, this being the first attempt of the kind ever made. Reports were given *in extenso* in all the London and many of the provincial papers, and I was informed by the chef some weeks after the dinner took place that he had received letters from Paris, New York, St. Petersburg, and many other Continental cities asking for information as to how to cook kid in the various ways represented. So great, in fact, was the number of his correspondents that he said he had serious thoughts of bringing out a book on the subject.

Fawn was not represented in the above menu, although this may easily be imitated with kid.

Whilst on the subject of the cuisine I may as well give here a recipe for cooking kid in this fashion, which, although elaborate and costly, some epicurean reader may be glad to try—if he can only induce his cook to follow the instructions here given.

Fattening and Killing

Neither goats nor kids as a rule are easily fattened, though there are occasionally exceptions, and I have myself seen a wheel-barrowful of suet taken from a she-goat of four years old. To fatten a sucking kid there is nothing equal to its natural milk, and if this is insufficiently supplied from its own dam, let it have in addition the milk of another goat. For kids that are weaned, grass during the day, with oats and bran morning and night, are the best kinds of food to get them into a plump condition. Castration of the male, as I have already observed, and "spaying" of the female contribute very materially to the secretion of fat, rendering goats fit to kill in a much shorter space of time. These animals are more easily fattened in summer than in winter if suitable pasturage can be supplied. Beans and peas, or the meal of these, mixed with some Thorley's Food and added to the oats or barley, will hasten the process ; but the taste of these creatures differs so considerably that it is impossible to say what dry food is best to give to lay on flesh generally.

The best way to kill a goat or kid is to take a long-pointed knife and thrust it into the neck just behind the jawbone ; the incision should be larger on the side where the knife comes out than where it goes in, in order that the blood may flow freely without spurting out. In this manner life becomes extinct in a very short space of time. The animal should have fasted for twenty-four hours previous to being slaughtered, in order to clear the intestines.

CHAPTER XX

OTHER PRODUCTS OF THE GOAT

Goats' Hair

Besides the milk and flesh, which in England and on the Continent are the chief objects of keeping the goat, the hair and skin possess commercial value.

The hair of the common varieties is manufactured in some countries into a strong though coarse kind of fabric, and worn as garments by the peasantry. That of the shaggy kinds, especially the males, is employed for making barristers' and judges' wigs. It is also excellent for ropes that are to be used in water, as they last a considerably longer time than those made with hemp. The hair may be shorn annually about the middle of May, in the same way as the wool from the sheep.

Many goats have besides hair a very fine fleece, which on some individuals is very plentiful, coming off in the spring upon everything that the creature's coat touches. Although certainly not as valuable as ordinary wool, it is by no means useless, but the quantity on a single animal is, comparatively speaking, so small as to be scarcely worth collecting. In Russia it has long been employed for articles of dress, such as gloves, stockings, etc., and is highly valued. It is stated that on one occasion some of this wool was woven with silk into a very beautiful shawl, the texture of which was greatly admired.

Mohair and its Use in Commerce.—Mohair, the name by which the fine silky hair of the Angora is known in the trade, must be regarded commercially as the most important product of the goat, employing as it does in this and other countries many thousands of persons in its manufacture.

Mohair first became an article of commerce in 1749, when a branch of the Levant Company, consisting of a few Dutch and English merchants, settled in or near the town of Angora, and commenced buying mohair and spinning yarns for export; the export of the article in its raw and unmanufactured state being then prohibited by the Turkish Government.

Tournefort, writing on the Angora goat in 1655, says that at this early period mohair yarn was spun into textures at Brussels, and in England was used in the making up of periwigs. Goods made of mohair were not known commercially in England, however, until early in the eighteenth century. It was about the year 1836 that the spinning of mohair became an industry in this country. The story goes that "a young man wandering about the docks at Liverpool was attracted by a quantity of long-fibred, frowsy stuff, the like of which he had never seen and the use of which nobody seemed to know. It had come from South America, 300 bales of it, and had been lying in the warehouse for months without a purchaser. The next day he returned and offered 1s. 6d. per lb. for the whole lot, an offer which was accepted with alacrity. The stuff was alpaca, and the young man afterwards became Sir Titus Salt." This gave the idea of spinning mohair, which for some years was classed commercially with the produce of the alpaca and the Cashmere goat as "goats' wool." The industry rapidly developed, and in 1853 Titus Salt began erecting the extensive works at Bradford which are now probably the largest in the world devoted to this manufacture.

Dr. J. L. Hayes (author of "The Angora Goat: Its Origin, Culture, and Products," 1882) says: "Mohair is not a substitute for wool, but occupies its own place in the textile fabrics. It has the aspect, feel, and lustre of silk without its suppleness. It differs materially from wool in the want of the felting quality, so that the stuffs made of it have the fibres distinctly separated and are always brilliant. They do not retain the dust or spots, and are thus particularly valuable for furniture goods. The fibre is dyed with great facility, and is the only textile fibre which takes equally the dyes destined for all tissues. On account of the stiffness of the fibre it is rarely woven alone; that is, when used for the filling, the warp is usually of cotton, silk, or wool—and the reverse. It is not desired for its softness in addition to silkiness—such qualities as are found in Cashmere and Mauchamp wool—but for the elasticity, lustre, and durability of the fibre, with sufficient fineness to enable it to be spun. . . Its lustre and durability peculiarly fit this material for the manufacture of braids, buttons, and bindings, which greatly outwear those of silk and wool."

The chief use to which mohair is applied consists in the manufacture of Utrecht velvets, generally known as furniture plush, and largely used in France, Germany, and the United States for the linings of railway carriages, also for sofas, chairs and table-covers. A great proportion of this Utrecht velvet is made at Amiens, in France, the mohair being previously spun at Bradford, the centre of the mohair trade in England, which, indeed, supplies the mohair yarn for the whole of Europe. Of late years Angora wool has been extensively used in the manufacture of ladies' light dress goods, known as "brilliantines," or "lustres," producing a fabric which, as the names imply, has a peculiar silky or lustrous appearance. Latterly, however, the fashion having gone out for these materials—for a time only probably—the trade has greatly diminished, and the price of mohair has depreciated in consequence. Amongst other manufactures which owe their origin to mohair may be mentioned the imitation sealskins which at one time were so much worn. To such perfection have these goods reached that it would be difficult without a close inspection to distinguish the imitation sealskin from the genuine article.

Another use to which mohair is applied, and that to a considerable extent, is for dolls' hair, the fine, soft, glossy nature of the article rendering it well adapted for this purpose. Some thousands of pounds sterling are annually spent in providing the *chevelure* for these toy babies. Finally, travelling rugs, made to imitate lion and tiger skins, astrachans, niggerheads, and every variety of imitation skins, are all produced from the wool of the Angora goat.

GOAT-SKINS

The operation of flaying should be performed as soon after the death of the animal as possible, for if it be delayed any length of time the hide may deteriorate in quality; this is sure to be the case if the goat died from disease and has been left till decomposition has begun to take place. To remove a skin properly requires some skill and care, so as not to cut it with the knife, and at the same time to leave as little flesh and fat attached to it as possible. Those who are inexperienced in such work had better employ their butcher's slaughterman, who for a trifle will kill, flay, and cut up their goat in a workmanlike manner.

When the skin has been taken off, all the bits of flesh and fat adhering to it should be carefully removed with a knife, and the hide placed to dry, the hair side inward, in a covered airy place free from damp. To prevent it from shrinking, the head and tail ends should be stretched out and nailed on a board, and the leg parts spread out with skewers.

Skins are sometimes preserved with salt and dried afterwards, but salt should not be used where it is intended to convert them subsequently into leather, as it never becomes thoroughly eradicated. The process of salting consists of laying the skins flat on the ground and well sprinkling the flesh side with salt and alum, more particularly on the edges and spinal portions. They are then folded by being doubled, first lengthwise down the centre, and then one fold over the other until a square is formed ; they will keep good in this manner for a considerable time, and may be dried afterwards. Any furrier would dress the skin for a small sum.

Goat-skins tanned and transformed into leather are valuable for the uppers of boots and shoes, being soft, elastic, durable, and damp-resisting. To procure the genuine article is, I believe, by no means easy, and when obtainable it fetches a high price, being much dearer than ordinary calf, or what is commonly called kid. Persons keeping their own goats have therefore an advantage in this respect. The difficulty is to get the skins tanned, most tanners refusing to take in the small number that a private individual has ready at a time. As for tanning them oneself, it is entirely out of the question, the process being varied and complicated, requiring at the same time a considerable amount of skill ; besides which the cost of materials would, for the few skins to be tanned, be considerably greater than their ultimate value. Dressing them roughly with the hair on, for mats and such purposes, is, however, easier, and may be effected as follows :

First, the skins should be soaked in water for about eighteen hours, to get rid of all dirt and blood, frequently working them about, more thoroughly to cleanse and soften them. They should next be scraped with a blunt, thick-bladed knife and thinned on the flesh side in order to remove as much as possible those portions of animal matter which are liable to putrefaction. After this they should be steeped for ten or fifteen days in a fermenting mixture of bran and water, composed of two pounds of bran to a gallon of water. On being taken out, they must again be washed, then folded with the hair sides in contact and immersed for ten minutes in a solution of alum and salt, in the proportion of a pound and a half of alum and a quarter of a pound of salt to a gallon of water. The quantities given are sufficient for eight skins.

The skins are then laid out and spread on the flesh side with a paste made by adding gradually to the last mixture, while constantly stirring, first a pound and a half of wheat flour, and subsequently the yolks of half a dozen eggs, afterwards incorporating the whole together. This paste has a softening effect upon the skins, making them also white, and counteracting any tendency to brittleness. They must be left in this state for eighteen or twenty hours, when, becoming stiff, they must be again rinsed in clean water, and dried by being stretched upon poles and exposed in a dry loft, where they must be left for a week or more as may be necessary. A good finish may be given by polishing the skins with pumice, rubbed on as hard as possible, and finally ironing over with a smooth flat-iron carefully heated.

Although this process effectually tans and preserves the skins, they cannot, of course, be expected to have that even, soft, and highly-finished appearance which those possess that are worked by a professional tanner. These effects are produced by a variety of instruments and tools which an amateur does not possess, and which it would scarcely pay him to procure, such as fleshing-knives, softening and stretching irons, beams, etc. Skins of goats with long hair may be improved, after the first washing and rinsing process has been gone through, by laying them out on a table or board and carding the hair with a coarse wool comb, paying particular attention to any dirty spots, which should be washed with soap. The skin of an Angora goat makes, when tanned, a very handsome mat.

Goat-skins vary greatly in quality, according to the breed of the animal, those coming from cold climates being better, and consequently more valuable, than those from southern latitudes. They are therefore adapted to different purposes; for instance, the hides of the Cape goats are most suitable for boot uppers; those of the Angora for rugs, muffs, and trimmings; whilst those of the English and Irish varieties, although less valuable than the kinds previously mentioned on account of being more domesticated, are nevertheless also useful for boots, furniture, purses, and coachmakers' purposes. In consequence of goat-skins taking a better dye than the skins of other animals, they are used principally for the manufacture of morocco leather, for which purpose they are tanned and dyed in a peculiar manner. It is in the form of morocco that such skins are employed for most of the purposes just alluded to. The skins of kids, on account of their clear and beautiful texture, are peculiarly adapted for gloves and ladies' boots, and in some parts of the Continent are even made into stockings and shirts.

Horns and Suet

The horns of the goat have been converted by some people into handles for knives and such like tools. The suet, which is generally very abundant in a full-grown and well-conditioned specimen, is said to be vastly superior in whiteness and quality to that of any other animal.

CHAPTER XXI

THE GOAT AS A FOSTER-MOTHER

A GOAT may be induced to bring up the kids of other goats after she has weaned her own, though these animals will not do so as readily as cows. There is no difficulty in getting a goat to take to other kids when they are put to her soon after they are born, especially if she has been recently delivered of her own; but when the little strangers are brought after the others have been weaned and are removed, she generally refuses to allow them to suck, and not infrequently ill-treats them. Some goats are more unmanageable in this respect than others, but when great repugnance to the new-comers is shown the following method may be adopted, and will be found in most cases to be effective:

Place the goat it is intended to use as a foster-parent by herself for a few days, and let her udder get each time well distended before putting the kids to her. Goats, like most other animals, dislike solitary confinement, preferring society, especially of their kind. They will therefore more readily associate with the strange kids when there are no others to fraternise with. By allowing the udder to get distended, discomfort to the goat is produced, which is relieved as soon as the kids begin to suck, and this, in conjunction with the advantage already gained, will often lead to the goat adopting the new kids as her own. These should not, however, be left with her the first time, but removed as soon as they have had a sufficient quantity of milk, the goat being held by the horns or head-stall while the kids suck, and only released and allowed to stand free when she no longer shows any opposition. Should any milk be left by the kids it must of course be drawn by hand. After a few attempts it will be noticed that the goat looks forward to the appearance of the kids to relieve her of her milk, and by the maternal expression of her face when they are sucking will readily show that her feelings towards them have undergone a change, after which they may be left together.

A goat will occasionally take to other kids willingly, and of her own accord, though such cases are not common. A remarkable instance of the kind came under my observation some years ago, which shows that these animals have their sympathies like human beings. This goat which belonged to Mr. B. Ravenscroft, of St. Albans, had given birth to two kids; but the maternal instinct in the animal seemed so imperfectly developed that she not only refused to allow her young to suck, but even maltreated them whenever they approached her. Another goat in the same stable, which had kidded some eight weeks previously, but whose kids had died a few days after birth seemed to take pity on these ill-used little ones, and by her sympathetic bleatings encouraged them to come to her for the nourishment their own dam refused to give them. And not only this, but she evidenced the disgust she felt at the unmotherly conduct of the other goat by punishing her with her horns whenever the latter attempted to butt at the kids. The strangest part of the affair, however, consisted in the fact that this benevolently-disposed animal was out of condition at the time, and yielding scarcely any milk, so that it could not have been from any selfish desire to be relieved of the pressure of an accumulation in her udder that led her to act in this manner. On the contrary, from the way in which these hungry little creatures tugged at her teats, with the incessant bobbing up of their heads to induce a better flow, the process must have caused pain rather than pleasure. Yet would she quietly stand chewing the cud, as satisfied and contented as though feeling happy from the knowledge that she was doing a kind action.

Goats are not infrequently used as foster-mothers for young animals other than their own species. I have myself had a calf partly reared in this way by letting it suck a large goat that gave me over three quarts of milk daily, and the calf thrived uncommonly well. The foster-mother in this case had to be held on a raised bench to enable the bovine offspring to get at her udder. A letter once appeared in the *Field* newspaper, with the writer's name and address appended, which stated that at Feering in Essex, a goat which had two years before produced two kids, after these were weaned, suckled first a young pig, afterwards two fawns, and finally two more fallow-deer fawns, which she was then fondly bringing up. All these several animals were successively fostered on her until they were weaned, and the milk flow was still maintained.

The most remarkable case I ever heard of a goat acting the part of foster-mother, and which was related to me on good authority, names, etc., being given, was as follows: "A gentleman of good position was left a widower with an infant child, her mother having died on giving her birth. He was of an eccentric turn of mind, and could not bear the idea of his child being nursed by another woman, and at the same time he strongly objected to bringing it up by hand with a feeding-bottle. In this dilemma it occurred to him to obtain a goat newly kidded, and to try and train the animal to suckle the infant after the fashion of the she-wolf in the story of Romulus and Remus. The experiment was successful, and the goat took readily to the child, running to it whenever it cried as if it were its own "kid." This went on until in due course, and in order to maintain the supply of milk, the goat had a fresh lot of kids, and here the strangest part of the anecdote has to be told. This extraordinary animal showed not the slightest affection for its new progeny, but entirely neglected them in its desire to supply nourishment to the child." When the tale was related to me, the child had nearly attained womanhood, being then in her seventeenth year, and the wonderful goat, I was informed, still lived.

CHAPTER XXII

HE-GOATS AND GOAT-CARRIAGES

I HAVE had occasion to refer to the male goat whilst writing on Breeding, but as this is not the only purpose that the animal serves I find it necessary to give him a chapter to himself.

The buck goat, vulgarly known as the "billy,"[1] is not a creature to be held in much esteem or to command any great degree of admiration, owing to two most objectionable attributes, viz., his disgusting habits, and the unpleasant smell he emits. The former I need not particularise; let it suffice to say that an entire goat is an animal which it is best, as a rule to keep at a distance. It is doubtless on this account that so comparatively few are met with. The peculiar odour emanates from the skin, and begins to be noticeable about the age of puberty, which may be any time after six months old, according to the precocity of the animal. This odour develops as the goat gets older, being always much more powerful at the rutting season than at any other time of the year. The scent may then in an old goat be distinguished half a mile distant.

He-Goats with Horses

Although unpleasant to the olfactory organs, this odour is by no means unhealthy, but rather the reverse with animals, especially with horses, on which it is supposed to act as a preventive against the staggers. For this reason he-goats are mostly kept by innkeepers and persons having large stables. I am well aware that some people regard this supposed connection between the hircine odour and the health of horses as ridiculous, and based simply on superstition; but for my part I

[1] It is high time that goat-keepers abandoned the almost childish terms "billy" and "nanny" and adopted the more sensible old English words "buck and "doe" in use in America. I am hoping that the B.G.S. will give official recognition to this suggestion.

cannot believe that an opinion which has been in favour for centuries with stock-owners, not only in England and the continent of Europe, but in lands still further distant, China and Japan, for instance, can after all have no foundation in fact.

Numerous are the cases that have been cited to me at various times of the beneficial effect of keeping a he-goat amongst horses, but the following being specially fixed upon my memory I give it here: My informant was a large contractor who, having lost annually several horses by the staggers, was one day advised to try keeping a "billy." He had always, he told me, scouted the idea that the smell of this animal, could in any way influence the health of a horse, but in despair determined to give it a trial, and bought one. It turned out that for three or four years whilst the goat lived not a single death from the old enemy occurred, and thinking he had eradicated the disease once and for all from the premises and was safe for the future, he did not at once replace the animal. To his dismay, however, in less than six months another death occurred in his stables from the same disorder. He then lost no time in procuring a fresh goat, and no more horses died from this cause.

Another case almost identical with this is recorded by Mr. Marshall in his work on the "Rural Economy of Gloucestershire," it having been related to him by Mr. William Peacy, of Northleach, whose authority he states is indisputable. Concerning this incident Mr. Marshall further observes:—"It appears to me probable that the influence of the goat is not merely that of a charm. The staggers appears evidently to be a nervous disorder. Odours are found in many cases, I believe, to act beneficially on the human nerves, and possibly the strong scent of the he-goat may have a similar effect on those of the horse. The subject is certainly entitled to enquiry."

The next case I have to relate came under my observation about the year 1877, when a serious epidemic which was very fatal in its effects raged amongst the breeding studs in Surrey, and generally around the metropolis. The Middle Park and Cobham Studs lost over thirty each, and at the Shepherd's Bush and Highfield Studs only four or five colts survived out of nineteen.

A gentleman kept at the time several brood mares, some of which he had recently bought from the Shepherd's Bush Stud. All foaled down well and not a single colt was affected, although his neighbours round were complaining of losses. In this case, however, a he-goat was kept with the horses, and the owner was firmly convinced that it was the smell of this animal—and it was strong enough, as I can myself testify—which kept his stock in health. As the gentleman mentioned was the only breeder who had goats and whose stud at the same time was free from the contagion, this case, combined with others, certainly goes far to prove that the popular idea is not altogether superstition.

Goats Preventing Abortion in Cattle

There is an old-fashioned and very prevalent idea that running a goat amongst cows is a preventive of abortion. I am well aware that the efficacy of this practice is also scouted and scoffed at by many people, and especially by members of the veterinary profession, who can see no possible explanation for it. As in the case of goats and horses, however, it has been found beneficial by so large a number of people all over the country, and instances are so numerous of the success that has attended the adoption of the practice, that I have no hesitation in giving space to the matter here. The following is one case amongst many of which I have personal knowledge. Mr. Sam Woodiwiss, whose name as a goat-keeper has been several times mentioned in these pages, breeds on his farm, amongst other stock, pedigree Red Polls. The farm proper, where the bulk of these cows are kept, is situate about half a mile from the house, while at the stud farm a few only are retained for domestic use. The goats are located in the latter establishment. Now, at the former, as at all his neighbours' farms, abortion is more or less prevalent, but amongst the cows kept at the stud no case has ever occurred during the nine or ten years that the place has been the present owner's property.

It had always been my impression that it was the scent of the *male* goat which had the supposed quality in question until one day I chanced to meet a large breeder and exhibitor of longhorn cattle—a great believer in the utility of the goat as a preventive of abortion—and he surprised me by saying it was always a she-goat that he had running with his herd. On one or two occasions when such a goat died and no other was available, abortion reappeared, but so long as a goat was there he was never troubled in that way. One of the theories advanced to account for this is that a goat pastured with cattle eats with impunity some kind of herb or grass which is the cause of abortion amongst cows, and if there be any truth in this it is easily understood that the sex of the goat is of no consideration. Anyway, the slipping of calves amongst a herd of cows is so general and its effect so disastrous to dairy farmers, whilst on the other hand to let a goat run with them is so simple and inexpensive a matter, that the only good argument that might be used against the efficacy of the plan is that if it were so invaluable it would be more generally practised.

EMASCULATION

It is a singular fact that by the process of emasculation the character and attributes of the male goat become completely changed. He is then as free from smell as the female, and no longer practises those offensive habits which render him when entire such an object of disgust, but becomes well-behaved, docile, and tractable, so that he may be petted and led about even by children. Under these circumstances, also, he fattens readily, and if killed when between one and three years old is really extremely good eating. Not only indeed are the inward qualities of the animal altered by castration, but his outer characteristics also undergo considerable modification. His horns and beard do not attain the same proportions, and his coat is often quite short, whilst he loses a good part of his masculine appearance.

All male kids not intended for stock purposes should be castrated as soon as possible, the best time being between two and three months old, and before the rutting season has commenced; it may be done later, but is then attended with some risk. At this age, on the contrary, no ill effects accrue, and although the creature becomes remarkably quiet for a day or two, lies down a good deal, and does not feed so freely, he soon regains his vivacity and former appetite. If older than three months it is better that the goat should be kept apart from the rest, especially from those of the opposite sex, for a week or so after ; and it is also advisable that the animal should have fasted for ten or twelve hours previously. Although the operation is a very simple one and easily performed, it should not be attempted by an unpractised hand. The services of a shepherd, or someone accustomed to operating on lambs or pigs, can generally be obtained, but it is better to employ a veterinary surgeon.

Goat-Carriages

He-goats that have been castrated are, when full-grown, very useful for drawing carriages, and as they possess great strength they are well-fitted for the purpose. The goats that are seen at seaside resorts, however, are generally either females —which are worthy of a better treatment and should not be put to such a use—or else they are under-sized, half-grown, wretched specimens of the other sex that are quite unequal to the task they have to perform, being objects of pity rather than admiration.

All goats may easily be taught carriage work, and will learn to obey the voice and rein as well as ponies. Their education should, however, be commenced at an early age, in order to get them thoroughly under control. The two principal requisites in their training are patience and kindness. These creatures are capricious in the extreme, at times appearing to refuse to perform what is required of them through pure whimsicality, which is better remedied by kindness and coaxing than by any application of the whip, as the latter only frightens them and makes them less tractable afterwards.

The trainer must begin by making his goat lead well, putting a bit in its mouth and the harness on its back, but not attaching it to any vehicle. This is, perhaps, the most troublesome and difficult part of the undertaking, as the animal does not at first quite understand what is required of him, and moves in any direction but the one in which he is required to go. The difficulty is overcome by practice, however, and the lesson should be gone through twice daily for an hour at a time.

When the goat has acquired tolerably well the habit of leading straight, and turning to right or left by the guide of the rein, it may be harnessed to the carriage, which should not at first be too heavy, and, of course, have no occupants, the trainer beginning by leading it about, and subsequently driving it with the reins, himself walking beside or behind the vehicle. It is a good plan, where a large open space is available to accustom the animal to the carriage and harness and to move freely by letting it go at will with the bit on and its head reined up ; this also gives it a better mouth for driving.

If a goat be exercised only once a day, this should be done in the morning before it has eaten any grass, or when it has been only lightly fed on chaff and corn, for some of these animals get so "blown" with green food that it is almost impossible to buckle the straps round them. They also go better on a half-empty stomach than on a very full one, besides which, if worked in the latter state, it spoils their "wind." Domestic goats are not long-winded at the best of times, and should, consequently, not be driven fast until in good working condition, and even then not for any length of time.

The harness should consist, besides the reins, of a bridle with bit, a breast-strap, a pad or saddle, with crupper, belly-strap, and shaft-tug to support the shafts. Blinkers are not necessary, though when used they improve the appearance of the little turnout, as does also a nice saddle-cloth under the pad. A small collar may be used in lieu of a breast-strap, and looks more finished, but the latter is easier and quicker to put on, besides being less expensive. A set of harness costs from £3 to £5, according to the amount and quality of the trappings. A good four-wheeled goat-carriage may be had for about £7.

A Nubian or half-bred Nubian goat harnessed in a well-made cart is a most attractive turnout. This breed is particularly suitable for the purpose owing to its great size, close glossy coat, long legs, and small horns ; but for the last-named and the different kind of tail it would resemble a miniature horse.

Some female goats have strongly pronounced male attributes, and may never prove reproductive. Such goats, however, grow to an unusual size and, though useless for breeding, make capital animals for draught.

Goats and Dogs

Goats and dogs are not naturally amicable, but they are both sociable animals, and when reared together they enjoy each others' companionship and often become fast friends. I have kept several varieties of dogs, and most of them took kindly to the goats, but a Sussex spaniel I once possessed had a most extraordinary attachment to his caprine associates. A she-goat that was trained to draw a carriage was his particular favourite, probably on account of the rides she gave him. The moment the goat was harnessed this dog would jump into the vehicle and thoroughly enjoy the ride he obtained, keeping his seat no matter at what pace the goat was going. When another dog was placed in the cart his jealousy and anger were unbounded, so that more than once they "had words," which led to open hostilities.

In South America, where large dogs are employed to guard the numerous herds of goats from the depredations of wolves and other carnivora, it is customary to rear a certain number of puppies under the goats. The natural affection thus engendered between the foster-mothers and their adopted offspring causes the dogs to live entirely with the goats, and thus makes them more effective as guards to the herds.

Taking Photographs of Goats

A large proportion of goat-owners are amateur photographers and are fond of taking pictures of their best specimens, not infrequently with the idea of illustrating an article. It is seldom, however, that these productions of the camera are suitable for that purpose, and the owner of the goat is disappointed on receiving back his print with an expression of regret that it cannot be utilised. Even when a photograph can be reproduced the effect is often unsatisfactory. One very common fault is to pose the goat with the head facing the operator, with the result that this part of the animal is far out of proportion to the size of the body, giving it a sort of "nightmare" character. Another is that the person holding the goat—who generally comes out much better than the goat itself—places himself in such a position that part of the animal, and often the most important, is merged more or less into the clothing of the attendant and the outline lost.

The goat should be posed broadside to the camera, with both head and tail an equal distance from the lens.

The camera should be set low down, so that the lens is about level with the middle of the body. Focus as sharply as possible, and develop for a strong negative.

In taking a rear view of a goat to show the udder, it is especially important to lower the camera to a level with the udder.

CHAPTER XXIII

GOAT-FARMING

In the Mont d'Or district in France, in Switzerland, and also in Norway, goat-farming has been carried out successfully, though under small proprietaries. In the first-mentioned locality, to the north-west of Lyons, some ten thousand goats were at one time kept constantly housed and stall-fed. This neighbourhood is very thickly inhabited, the ground being apportioned amongst smallholders, whose land is so closely cultivated that pasturing is out of the question, and the soiling system has to be adopted. They are thus, as regards facilities for pasturing, in a still worse position than we are in England, where at least cheap grazing can often be obtained, though we have not advantages for turning out stock to wander about at will and find their own food as have the mountaineers of Norway, Switzerland, and some other countries.

At the same time, these small farmers of the Mont d'Or received far lower prices for their produce than can be obtained in England. These people kept from twenty-five to thirty goats, each of which was reckoned to yield about 600 litres, or 528 quarts, a year. The milk was used almost entirely in the manufacture of the famed Mont d'Or cheese, well known and appreciated throughout France, though the sale has considerably fallen off, I believe, of late years. The year's supply of milk from one goat was reckoned to produce 578 cheeses, which sold for 20 centimes, or 2d. each, thus placing the price per litre (1¾ pints) at a fraction over 2d., or little more than 2¼d. per quart. The kids were sold at a fortnight or three weeks old at 3 francs each, so that, allowing a couple to every goat, the gross return from each animal was 130 francs, or in English money about £5 3s. The cost, on the other hand, for food, wages, etc., allowing for depreciation in value of stock and interest on capital sunk, was set down at £3 5s., showing a net profit per goat of £1 18s.

An Experimental Goat Farm

A goat dairy farm was started in 1882 on an eminence among the Surrey hills, between Leatherhead and Dorking. I paid a visit to this farm shortly after it was established, there being then about 120 goats of a very mixed and nondescript character, most of the animals of the long-haired Irish breed. The experiment was not a success under the management of the original proprietor, who was unable to obtain a ready sale all at once for the quantity of milk he had, and the farm was subsequently sold to the Express Dairy Company, in whose hands, had success been at all possible, it would no doubt have proved a profitable undertaking. There were, however, many difficulties to contend with, the chief being that of securing a sufficient number of goats in milk to supply the demand in autumn and winter, and of creating a demand equal to the supply in summer. As a result the farm after a short time was given up as unprofitable.

Is Goat-Farming Profitable ?

Every now and again some inexperienced person writes to the Press advocating goat-farming, and not many years ago a lady devoted a series of articles to this topic with a view to showing that a small fortune was to be made out of such a business. The writer seemed to think that because a few goats kept privately to supply a household with milk are a profitable investment, a large herd the milk of which was sold wholesale as is done with cows would be an equally remunerative undertaking. Possibly it might, but the first requirement in such a case would be to ensure a ready sale for the milk, and up to the present there is not the slightest reason to suppose that there exists commercially a demand for this article. People who require goats' milk wish to have it from their own goats, just as most householders would like their cow's milk from their own cow if they had the means and the facilities for keeping one. With a goat this is, of course, a much easier matter; the consequence is that there is a very considerable demand for goats, but very little for the milk.

The question therefore arises: Would it pay to farm goats to supply this demand? The answer is extremely doubtful, and so far there is little evidence in support of a favourable reply. It all depends on the prices obtainable for the goats sold. On this point perhaps no one is better able to express an opinion than Mr. Sam Woodiwiss, who writes as follows:

"In reply to your question as to whether I have found goat-farming profitable without selling any milk, I should state that until I gave up breeding for size and form only and made a speciality of milk there was certainly no profit in it. This may appear inconsistent considering that the sale of milk forms no part of the returns, but the explanation lies in the fact that the prices I realised under the old system when I kept only Anglo-Nubians were insufficient to show a profit. I found that people who applied to me for goats in a general way wanted animals giving from two to three quarts daily, many expecting to obtain them for £2 or £3 apiece. What few I had giving those quantities I deemed it advisable to keep to breed from. When I took up with Toggenburgs and Alpines and won milking prizes with them I soon found breeders anxious for my stock and ready to pay remunerative prices.

"It should be remarked that in establishing my herd of milkers I spared no expense to obtain the best stock that money could buy. Amongst these were, as you know, the Alpine goats imported from France, some coming from the Paris show and others from M. Crepin's herd, as well as a fine male Toggenburg from the Jardin d'Acclimatation. I thus established a strain of milking-goats, a few of which I mated with Anglo-Nubians, but preferred to keep, as far as possible, to what I called, in a general way, 'Swiss' blood.

I can therefore truly say that under present conditions of all the stock on my farm, excepting, perhaps, my herd of dairy Red Polls, the goats are the most profitable. This may be judged by a reference to the prices obtained. In the course of the last six years I have sold 103 goats to the value of £721 2s. 6d., *i.e.*, about £7 a head on an average, reckoning goats and kids:—

"On the debit side the expenses are low. I reckon each

full-grown goat costs me during the five summer months on an average 6d. a week to keep, and during the seven winter months about 1s. a week. It must be borne in mind that I grow my own hay and roots and thresh my own oats (the only corn I use, and that only in winter). During summer I find my goats milk better on pasturage and hedge-trimmings. My cost of feeding would thus be less than half of that reckoned by yourself as the expense for food when everything except garden produce has to be bought. Other items of expenditure are very trifling, except that of sending to shows and exhibiting, but as I do not put down the prize-money obtained, which has amounted to a good deal in the course of a season, I do not take these items into consideration.

"On the whole I may say I find goat-farming now decidedly profitable."

CHAPTER XXIV

SHOWS AND SHOWING

THE first goat show ever held in England took place in connection with the Newton Abbot meeting of the Devon County Agricultural Association on 19th May, 1875. The prizes were given by the Baroness Burdett-Coutts, who was staying in the neighbourhood at the time, and at whose instigation the classes were formed. These were divided into short-haired, long-haired, medium-haired, and hornless goats, with a class for kids under six months old. In the summer of the same year an important show of goats took place at the Crystal Palace, when as many as 108 entries were made, the prize list amounting to £72.

In October, 1876, the year which saw the inauguration of Dairy Shows at Islington, goats formed part of the exhibits, and have continued doing so annually without interruption to the present time. Being held in a permanent building, and at a time of year when goats are not generally in full milk, these classes are largely patronised by the principal breeders, so that nowhere is there such a fine collection to be seen as at these gatherings. He-goats, however, whether adults or kids, are not admitted here on account of their objectionable odour, for, although for many years male kids up to nine months of age were allowed to be entered, one precocious youngster even at that age, it was found, would taint the whole section and bring the exhibits generally into disrepute, and so he-goats were eliminated from the schedule.

Two subsequent shows to the one mentioned in 1875 have been held at the Crystal Palace, viz., in 1876 and 1883, the latter being under the direction of the British Goat Society, and these were followed by classes for goats in connection with mules and donkeys at the Alexandra Palace in 1880 and 1881. At Kilburn in 1879 the first goat classes were included in the Royal Agricultural Society's schedule, and these were repeated at Windsor in 1889, and at Manchester in 1898.

Shows have not been held annually by the B.G.S. since, as so many county societies now include these classes. From 1896 the Tunbridge Wells show has never been without this feature in its catalogue of exhibits. These competitions have probably contributed more than anything else to bring goats into public notice and to improve the breeds for milk, as at most shows at the present time milking trials are included. A vast deal of good might still be done towards both developing the milking qualities of these animals and rendering them more popular if agricultural societies throughout the country would imitate the example set by those just mentioned, and include classes for goats in their schedules. Although the entries may be few on the first occasion, they generally increase year by year, and the exhibits always form an attractive feature, owing probably to their novelty and to the interest attaching to this miniature dairy stock.

Hints on Arranging a Show

As persons desirous of getting up local shows are generally glad of information from those who have had more experience in such matters than themselves, and as in the case of goats there are several important details which, if omitted, often lead to unsatisfactory results, it will probably be of service if I offer a few hints on this subject. Goat-fanciers are often reluctant to enter their stock, partly because these exhibitions are not such everyday occurrences as those of dogs and poultry, and it is not therefore so fashionable to do so, and partly because, the points not being well known, many people think their stock not good enough. It is therefore necessary, before deciding on the show, to communicate—personally for preference—with all the breeders of goats in the neighbourhood, and induce them to take part in the competition. Here I may mention that the British Goat Society is generally willing to assist in the promotion of such shows by offering a medal as a special prize, and in giving advice towards carrying out the arrangements.

Schedules

It is at present, and will doubtless continue to be for some years, impossible to arrange the classification according to breeds, chiefly because there are so few pure specimens of the different varieties, and such a very small proportion of goat-keepers conversant with them. It was usual in the early days of shows to divide the classes according to length of hair, arranging them as "short-haired" and "long-haired" goats. The former generally embraced the English and Anglo-Nubian breeds, whilst the long-haired specimens were represented by Irish and Welsh goats. The objection to this was that so many exhibits were not strictly either one or the other, and some had long hair on the ridge of the back and on the thighs and buttocks, while it was short on other parts of the body. In addition to this a long coat soon became unpopular with British fanciers, so this system of classification was abandoned giving place to "horned" and "hornless" goats which, is a well-defined distinction, though meaning nothing in the way of breed.

It has also been found necessary for some years to make a separate class for "Toggenburgs and other Swiss breeds," and in the larger shows to separate goats that have won "two or more" or "three or more" first prizes from those which have been less successful in the prize ring, in order to give these latter a better chance.

In providing a class for kids it is a mistake to require that they should be "in pairs," which, strictly speaking, implies male and female. The objection to this is that an exhibitor who may possess a very good doe kid has an inferior buck, or perhaps none at all, to send with it, and consequently he either does not show his one good animal, or by sending both spoils his chance with the indifferent specimen. If the class comprises male and female kids it is generally unsatisfactory both to the exhibitor and to the judge. It is far better, therefore, to provide separate classes for each sex amongst kids as well as goats.

A class for goatlings—*i.e.*, she-goats over one year old and under two years—should be provided if possible, with a condition of entry that any exhibit that has borne a kid will be disqualified. Since the establishment of such classes the effect has been most marked in obtaining goats of increased size and bulk, in consequence of breeding being deferred until a perfect development of the frame had been secured. The condition at one time was not only that the goatling should never have *bred*, but that it should never have been *mated*. In consequence, however, of the difficulty of proving whether the animal is in young or not, supposing it to have been served within a month or so of the show, with the objection also that themating may be thus unduly protracted, the maiden condition is not now required to be observed. There is this disadvantage, however, in allowing maiden goatlings and those that are in kid to be exhibited together, that the latter, by reason of their extra rotundity and the expansion of the ribs due to pregnancy, present a better appearance to the judge, although as a matter of fact, they may not be really superior to the others.

When the amount of money at the disposal of a committee admits of it, prizes should be offered in a separate class for the goats yielding the largest quantity of milk, the class being headed "Milking Class" to distinguish it from the inspection classes. As here productiveness is the one point to consider, it gives a chance to exhibitors who have animals with no pretensions to size, quality, breed or other attributes pleasing to the eye, but which simply fill the pail. In such cases, however, the goats have to be all milked out dry in the presence of a steward or other show official the evening before the show opens—generally at 7 p.m. This is a practical matter, the awards being based upon actual facts and not on the opinion or fancy of a judge.

To attract entries there should be, if possible, three prizes offered in each class, and these not of less value than £1 10s., £1, and 10s. respectively. Even then the cost of carriage to a show is so considerable when exhibits are sent any distance that the winning of a prize seldom covers the expense. The entry fee when such prizes are offered is generally 5s.

Benching, etc.

Goats never look so attractive at shows as when exhibited on raised benches in the same way as dogs. This, of course, comes rather expensive, and when the show is an open-air one in a field, the goats are generally placed on the ground and divided by hurdles. If benches, however, are used, it will be rather cheaper to have them double, that is, a double breadth, 6ft. wide, divided down the middle by boards fixed horizontally. For full-grown animals the amount of space allotted to each must be at least 3½ft.; but 4ft. is better if it can be given, especially to he-goats. For kids 2½ft. is sufficient. In all cases posts should be inserted, or laths of wood nailed over each stall, on which to fix the catalogue numbers of the exhibits and the prize-cards. These must be placed quite 6ft. from the ground or bench, otherwise the goats, by standing on their hind-legs, will eat them as fast as they are put up. This mischievous propensity has to be guarded against also in connection with the address-labels for travelling. The latter should have a string at both ends, so that they may be attached to the horns or round the collar, and not left hanging to the latter. Even then, if the goats get together one will often demolish the label of the other, and nothing gives the secretary of a show more trouble than having a lot of goats at the close of an exhibition to send back to their owners by rail with nothing to indicate to whom they belong. To avoid any such predicament it is necessary to furnish each exhibitor with a metal tablet stamped with the number of his goat, according to the catalogue, which should be fastened to the collar of the animal by a wire. These metal tallies can be procured from Spratt's Patent Co., Ltd. Another point to be observed is not to place the rings or staples at too great an elevation from the ground. About 15in. is the best height.

Feeding at Shows

Besides hay, it is necessary that some green food should be provided for the goats, in order to keep up the yield during the time the exhibits are under the care of the committee. To this end, if it is a summer or autumn show, arrangements should be made for a quantity of cut grass or tares to be supplied, or, if the season is too early or too late for this, a few roots. With the excitement and worry from the constant stream of exhibitors that are continually passing, goats do not consume much food at shows, and if the exhibition extends over several days the yield of the animals generally suffers in consequence. It is therefore very necessary that what food is supplied them should be of a succulent character.

Providing an Attendant

Exhibitors of stock such as horses and cattle invariably send a man in charge of them, but this is not always the case with goat exhibitors, many of whom only keep a few such animals, and would find it too expensive to incur the cost of a man to look after them. As goat-breeders have to be encouraged to exhibit rather than otherwise, it will be necessary for the committee of a show to provide a suitable man to take charge of the goats whilst at the show, and he should be able to milk as well as feed them. It is usual to pay 5s. a day for this service, the man being, of course, under the supervision of the steward, who will find it to his advantage to have such a person under him who can receive and fasten up the goats as they arrive on the ground, and replace them in their crates for delivery to the railway companies when the show is over. Such a man should be provided with a badge of some sort, so that he m ay be identified by the exhibitors.

Judging

When goat shows were first started, as the exhibits were then generally small the chief point to encourage was size, but when this—after some ten years' breeding under an improved system—was obtained, more attention was given to milk, and this is now the prevailing feature. The following is the order of precedence under which points are generally considered by judges at the present day: (1) Milking qualities; (2) size and shape; (3) quality and condition; (4) colour and markings. It will be, perhaps, of advantage if I discuss each of these several qualifications in detail under their respective headings:—

Milking Qualities.—It should perhaps be explained that in awarding prizes under inspection the actual amount of milk a goat gives at a "meal" is not taken into account: it is the appearance of the udder in conjunction with the length of time the animal has been in profit since the birth of its last kids. Nor is this all, for the mere appearance of the bag is not sufficient. It has to be handled to see if it is hard and fleshy or soft and pliable, and this sometimes cannot be ascertained until the goat has been milked. Lookers-on are thus apt to think, when they see the exhibit under inspection being milked out, that the judge is having this done to ascertain the amount yielded and not the state of the udder after the operation. Next to the udder come the teats, and these have to be considered in regard to their size, convenience of handling, their position in respect to the udder, and their evenness. Some goats have false teats, and these are more or less a defect. When large they come in the way of the milker, and if, as occasionally happens, milk issues from them, it is a great fault, and a really good milker would lose place considerably in consequence. To estimate properly the quality of a goat as a milker the judge should milk her himself, as in some animals the milk flows freely, whilst in others it is hard to obtain, and appears in a thin stream, which makes the process tedious and tiresome. Where the same goat, however, is entered in a milking class this cannot justly be done, or only to a trifling extent, as it means lessening the quantity of the yield when the test of productiveness is taken in pounds and ounces.

Size and Shape.—Size without shape goes for little, as shape means something more than symmetry or picturesqueness. The points embodied in the configuration of a goat have some reference, or at least are supposed to have some reference, to its milking qualities. These points are, in the main, a long, level body, with prominent withers and width between the hip bones, well-sprung ribs, a slender neck, a broad chest (indicating constitution rather than milk, however), and legs straight and of proportionate length. There should be a gradual deepening of the frame as it approaches the hindquarters, this being a decided milking attribute, but one which generally develops with age. The style of head depends very much on the breed, but there is such a thing as a "milking head," as it is often called, which is broad at the forehead but fine and tapering towards the muzzle, with little or preferably no beard—a delicate rather than a coarse head, in fact.

If there is a point on which judges are apt to give exercise to their fancy, it is in regard to horns. There is no doubt that most breeders, whether acting as judges or as buyers, give preference (other points being about equal) to a goat without horns; but if these ornaments, as some consider them, are present they should undoubtedly be small and slender rather than coarse and heavy. Sharp, pointed horns rising perpendicularly from the head are decidedly objectionable from a practical point of view, as they are unquestionably in some degree dangerous. The best-shaped, in the writer's opinion, are those which lie close to the head, curving back like the horns in some sheep. The former are generally noticeable in "Anglo-Nubians" having Indian blood in their veins.

Returning to the question of size and the remark made above that, *per se*, it carries but little recommendation, it should be here explained that no judge worthy the name would think of placing a tall, leggy goat, shallow of frame and with a small udder, over another, little more than half its height perhaps, showing a good shape and a large bag. Long legs give a gawky appearance to a goat. Short legs are no real fault, provided they are fairly proportionate to the body. Size is a quality affecting kids and goatlings rather than full-grown goats, but as we encourage it in these immature animals we cannot altogether ignore it when the same animals become fully developed.

QUALITY AND CONDITION.—These two points are bracketed together as they so often figure concurrently. The word "quality" may perhaps best be defined as the reverse of coarseness, *i.e.*, an animal symmetrically built, delicately shaped, fine in bone, with a close, glossy coat, soft hair, and a thin, supple skin. It is closely allied, too, to the question of health, for some of these features can hardly be present if the goat is unhealthy any more than they can if the animal is out of condition. Quality, in fact, is that peculiar attractiveness which is recognised and admired by everyone, whether the person be an expert on the subject or not, and it goes a long way towards bringing success to the exhibitor. Condition may mean the bloom of health or the "get up" for exhibition. It does not affect, we know, the intrinsic merit of an animal; but be that animal a dog, a cow, a horse, or a goat, no judge can afford to overlook it. Considering that in close competitions the qualities of some exhibits are so much on a par that it is almost impossible to say which is most deserving of a prize, a judge is always glad to have condition to fall back upon, as he knows his decision will then be generally approved. Although "quality" is a feature dependent in great measure on "breed," it is not always the case, as even in animals of the same breed it is more apparent in some individuals than in others. Moreover, the point in question can in great measure be artificially developed. A goat with a rough, ungroomed coat may have quality naturally, but in that state, it is not apparent. An experienced exhibitor knows how to put his animals in show form.

COLOUR AND MARKINGS.—These are matters which affect goats probably least of all exhibits, for the reason that, excepting Swiss varieties, this class of stock is not judged from the point of view of breed, which is so often governed by colour and marking. The remark made above in regard to horns applies here with still greater force. Colour is purely a matter of taste or fancy, and is only brought under consideration when all other qualities are equal. Some breeders admire a spotted goat, others a self-coloured one ; but in neither case has it any bearing on milk production. In dealing with Swiss goats colour plays a far more important part. Judges know by now the proper colour for a Toggenburg, a Saanen, or a Schwarzhals, if such a breed should be present ; but they would not be too particular perhaps—at the present day at any rate—as to the amount of white displayed in a Toggenburg, though they would hardly pass for one of the other varieties above mentioned a goat with the rich tan and black that so often indicate Nubian or Indian blood. Points of breed in the matter of colour are now officially defined.

JUDGING KIDS AND GOATLINGS.—As these, or the former at least, cannot well be judged on precisely the same lines as full-grown goats, something must be said here in special reference to them. Size in kids is one of the principal features, but only when taken in conjunction with age, for it is obvious that to award a prize to one animal because it is bigger than another without investigating the respective dates of birth would be grossly unfair. When goat-showing was in its infancy, and before the Kid Register was instituted, judges had to rely on the teeth of the animal to determine its age, and so a kid with two of its incisors changed was in those days disqualified, although perhaps the animal was really under twelve months old. This led to much unpleasantness, and was very awkward for judges. Now that every kid must have been registered to qualify for a prize, the adjudicator does not trouble to examine the teeth, but takes the ages as they are set before him in his judging-book. And here I would remark that no judge should start work on these classes until the steward or secretary has supplied him with this necessary information, if it is not already before him.

In judging quite young stock proportion or symmetry cannot be taken much into account, because the kid that is going to make a big goat is generally, like a puppy, of ungainly appearance, "all legs and wings," as it often is described. This is well shown in the illustration of Grey Rock, the largest male of his day. At the time the photograph was taken he was a kid of six months, and I well remember his ungainly, though promising, appearance when winning first prize in the kid class at Tunbridge Wells, in 1898, the first of his many honours. In a kid, length and depth of body, well-rounded ribs, plenty of bone, and a neat head are, as in a fully-grown goat, points which are sought for. A "pretty" kid, however pleasing to the eye, is not necessarily the sort that will make a big goat or a good milker. At the goatling stage one can begin to look for the points which go to fill the pail, and precocity in the matter of udder and teats is a good sign in that direction.

Here comes in the vexed question of whether a goatling should be allowed to be shown in milk or be disqualified if it has borne a kid. For the sake of fairness and judging all on an equal footing, the latter should perhaps be insisted on, but there is such a thing as a goatling showing a well-developed bag, and even giving milk, without ever having been mated, and it would be equally unfair towards the owner of such an animal to have it disqualified because it exhibits—abnormally perhaps—the very quality that shows are intended to promote and most breeders are anxious to secure. In the absence however, of any rule in the schedule to the contrary, it seems to me that a judge can hardly shut his eyes to such unmistakable evidence of a coming milker, and refuse to place it above others in which the same quality is not displayed.

Judges are, indeed, often placed in this difficult position when adjudicating in a class of goatlings where an exhibit is in milk as a consequence of having bred or is due shortly to kid, as such an animal, assuming these milking features to be prominently displayed, has, as previously remarked, a decided advantage over others. It is, nevertheless, scarcely possible to ignore the very points which, if the animal were a month or two older perhaps, and entered in an adult class, would ensure its taking a high position ; and yet it must be admitted that by giving such an animal priority it is a direct encouragement to early breeding. In my opinion, if a precocious goatling exhibited in milk is at all under-sized in consequence of its condition in this respect, its inferior size might weigh against its merits as a milker. If, on the contrary, the size, in proportion to age, is equal to the best grown in the class, the animal should have the benefit of its unquestionable milking attributes. The breeder, then, who forces on his young stock in the way of early breeding does so at some risk.

HERMAPHRODITES.—Finally, I should mention a matter which some judges are apt to disregard, but which is of the greatest importance—that is, prominent sexual characteristics. There are amongst goats a largely increasing number of what are called "hermaphrodites," creatures which seem to combine to a certain extent the sexual organs of both the male and the female. Such goats are generally females, but have a decidedly masculine appearance. A very striking instance of this occurred some years ago, when I was judging at Tunbridge Wells. There was but one class for kids, exhibits of both sexes being present, and I was separating the young bucks from the does, in the process of adjudicating upon their respective merits, when I noticed amongst the latter the usual sexual evidence of the male, so I promptly had it shifted to the other side. When I examined the kids on that side, and took a posterior view of them, I discovered a female among them, and at the same time was struck with the similarity of the animal to that I had previously sorted out. This led to a closer test being made, and then I discovered the extraordinary and abnormal condition of the sexual organs. By good luck I had as my steward a veterinary surgeon, and I accordingly asked him to bring his professional knowledge to bear on the point, and to examine the animal critically. This was done, and the greatest surprise was expressed by my colleague on the evidence before him, the case being, as he stated, quite unique according to his experience. Here is proof how necessary and important it is for judges to examine the sexual organs of the exhibits carefully, whether kids or goats. Breeders should make a similar investigation of their kids from time to time, from birth, and when these hermaphrodites are discovered they should be destroyed, for a she-kid with these male attributes will never breed, and a male with the attributes of the opposite sex is generally useless as a stock-getter.[1]

Since the above was written a very able article from the pen of Mr. C. J. Davies, a well-known naturalist, has been published in the *Veterinary Journal.*

Mr. Davies is of opinion that this malformation, which he finds very common amongst Swiss and crossbred Swiss goats, and scarcely ever among goats of Nubian blood, is due to inbreeding. Inquiries made among Swiss veterinary surgeons have elicited the fact that the phenomenon "is by no means uncommon among goats in that country." The statement is nevertheless added that "there appear, however, to be other factors which have not been cleared up which are to be blamed for it."

Hints for Exhibitors

What has been written above applies chiefly to the organisers of a show and to the judge, though incidentally it affords information also to the exhibitor. I now consider it expedient to offer a few remarks to the exhibitor in particular. The first matter of importance is carefully to read and follow the regulations in the schedule prepared for the guidance of those about to exhibit, who very often neglect to peruse them, and have to

[1]Three Toggenburgs imported by Mr. Paul Thomas in 1887 gave birth to seven kids between them, and in each case their was one kid showing the peculiarity mentioned.

pay the penalty afterwards. The next is to be extremely careful in filling up the entry forms, especially in regard to the number of the class applicable to each goat entered. It is surprising how very often disqualification occurs through this. and exhibitors will seldom believe it is their own oversight until they see their handwriting on the entry form. A third warning is not to omit to sign and date the entry form, and not to forward it to the secretary without the prescribed amount of fees, as no secretary will accept an entry unless this very necessary formality is observed. When kids are to be exhibited it will be necessary to ensure their being entered in the Kid Register of the British Goat Society, and the number of such entry should be inserted on the form. If time is short before entries close, some arrangement can generally be come to between the exhibitor, the secretary of the Society, and the secretary of the show for this number to be received after entries have closed.

Metal Tallies

At all important shows the secretary furnishes each exhibitor with a metal tablet bearing the catalogue number stamped thereon, in addition to the address label. This is most necessary, as otherwise when two or three goats are sent together a fine opportunity is afforded them of displaying their innate love of mischief by eating off each other's labels, and so confounding the officials as to ownership and destination. Exhibitors should be careful to see that the metal tablet or tally is attached (by the wire that accompanies it) to the goat's collar when dispatched from home, and not left until the goat arrives at the show. When an attendant is sent to a show with a lot of goats in a horse-box he generally puts labels and tallies in his pocket, to be placed on the animals when he arrives with them on the show-ground. This work is then often done in a hurry and carelessly, with the result that a goat gets placed under the wrong number, and so figures incorrectly in the catalogue, if it does not—as may also happen—get disqualified for a prize.

Crates and Addresses

Another important point, both for the sake of the exhibitor and of the show officials, is to see that the crate in which a goat is sent bears the owner's name and address, either indelibly marked thereon or with an extra address-label securely tacked upon it. The former is much the better, as the ink on a label may be obliterated by rain when the crate is left exposed in the open, as not infrequently happens.

Carriage by Rail

Exhibitors in sending their goats to a show by rail must be careful to prepay the freight, as secretaries generally refuse to discharge these liabilities, not caring to take the responsibility of getting repaid afterwards. There is usually some additional charge for cartage from the station to the show-ground, and in this exhibitors may often help each other by sharing the hire of a horse and cart for the purpose, and so diminishing the cost. If the distance to the show ground is not more than a mile, however, the charge by railway companies is generally 1s. per goat.

Prize-Cards and Rosettes

Most exhibitors are anxious to secure these trophies of success after the conclusion of a show, but unless they are present themselves, or have someone to represent them, they rarely obtain either prize-card or rosette, as secretaries and stewards will not undertake their transmission, except in a private way as a personal favour. It is well in such cases to get some other exhibitor to allow his own man in charge of his goats to perform this act of kindness.

Coats for Goats

Although at nearly all shows the goats are placed under canvas-roofed shedding, if the weather is wet and windy they are apt to be exposed to driving rain and cold. It is therefore advisable to provide a goat that has been kept much in stables and is highly valued, with some kind of coat properly fitted to it in the same way as is done with greyhounds. This will, of course, be removed at the time of judging by the man appointed by the committee to take charge of the exhibits, and a small tip will generally ensure its being replaced and returned on the animal in the event of the owner not being present to look after it himself.

CHAPTER XXV

THE BRITISH GOAT SOCIETY

As this book seeks to include everything interesting to goat-keepers, it would scarcely fulfil its purpose without a chapter dealing with the British Goat Society, an institution that has done so much in the last fifty years to encourage goat-keeping and to improve the goat. Such a society had been a pet scheme of the present writer's as far back as the early 'seventies, when the first edition of this book was written. No serious effort was made, however, to put the idea into practice until the Kilburn Show of the Royal Agricultural Society of England in June, 1879, when, with a view to securing the introduction of goat classes into the schedule, a prize fund was opened, and the names of some fifty subscribers were obtained, all of whom were willing afterwards to form a syndicate to further the interests of goat-breeders.

A meeting of those interested was called at the Dairy Show on the 16th of October, 1879, when it was formally resolved to establish a society to be called the British Goat Society, and a provisional Committee was elected. This was followed by the holding of a general meeting during the Cattle Show week two months later, on which occasion a president (the Earl of Rosslyn) and Committee were elected, and the hon. secretary and treasurer were appointed.

The objects of the institution were declared to be:

1. To circulate knowledge and general information upon goats with a view to counteracting the prejudice and ignorance which prevail in a great degree concerning these animals.

2. To extend and encourage the keeping of goats, particularly by cottagers, in order to increase the supply and con-

sumption of milk in rural districts, where this article is frequently unobtainable.

3. To improve the various breeds of goats, and especially to develop those qualities which are generally recognised and valued in milch stock.

The minimum subscription at that period was 5s., but in later years this sum was restricted to *bonâ fide* cottagers or artisans, all other annual members being required to pay 10s., and life members £10.

The late Baroness Burdett-Coutts, who had for some years previously taken a practical interest in goat-breeding, was approached to give her name as patroness, and very willingly consented. The prestige which the name of that noble lady gave to any institution with which she was connected had a marked effect in establishing a reputation for this small syndicate of goat-keepers, and it was through her aid that such members of the nobility as the Duke of Wellington, the Duke of Portland, and the Earl of Londesborough were obtained as presidents or vice-presidents later.

PRESIDENTS

The following is a list of the Presidents of the Society :—

1880-81. The Earl of Rosslyn.
1881-83. The second Duke of Wellington, K.G.
1883-85. General Burnaby.
1885-86. W. J. Evelyn, Esq.
1886-92. The Earl of Rosslyn (second time).
1892-99. Sir Humphrey de Trafford, Bart.
1899-1902. H. C. Stephens, Esq.
1902-3. Sir Henry Isaacs.
1903-5. The Baroness Burdett-Coutts.
1905-8. The Duchess of Hamilton and Brandon.
1908-9. Sir T. C. Dyke Acland, Bart.
1910-11. The Countess Bathurst.
1912-13. The Marchioness of Aberdeen.
1914-16. Sir Humphrey de Trafford, Bart.

Later Presidents have been Sir Thomas (afterwards) Lord Dewar, Lord Leverhulme, Earl Bathurst, Mrs. J. C. Straker, Mr. Reg. Pease (1927) Lady Forteviot (1928) and the present writer (1929).

The Herd Book and Prize Record

One of the first objects of the Society was the establishment of the Herd Book and Prize Record. The former was no easy matter, as goat fanciers had never attempted to breed on the lines of purity, first because at that period there were no pure breeds to work upon, and secondly because when fancy points are aimed at practical qualities are frequently lost sight of. The only principle on which it was found possible to construct a Herd Book was that of "like producing like," or, in other words the proved capability of an animal possessing good qualities to reproduce the same in its progeny. This capability could only be ascertained and officially recognised when the goat itself and one or other of its parents had won a prize. The basis of the Herd Book therefore was the Prize Record, a register of goats that had won prizes since the first show of any importance, viz., that held at the Crystal Palace in 1875. The mere fact of the goat itself winning would not qualify it for entry, as it might be that the qualities which gained it the award were a mere accident, or that the award itself was due in great measure to lack of competition. It was necessary that the qualities should be sufficiently fixed in the blood to be capable of transmission. The following are the conditions of entry:

General Section

14. When both sire and dam are entered in the Herd Book.

15. When three of its grandparents are entered in the Herd Book and the sire or dam of the fourth is so entered.

16. When three of its grandparents are entered in the Herd Book, and the fourth was mated to one of the said grandparents, which must be a "dagger" male, or a "Star" or "Q Star" female goat.

17. When the sire or dam is in the Herd Book, and the other has won a prize at a recognised show in an adult class open to general competition, or in a Milking Competition, or has won a Star Award in accordance with Regulation 83.

18. When the goat itself and both sire and dam are in the Prize Record.

19. When the goat's granddam on the side of the sire and either the goat's dam or the dam's dam have won a prize in a

milking competition recognised by the Society, or has won a Star Award in accordance with Regulation 83.

20. When one parent is an imported goat entered in the Herd Book, and the sire or dam of the other parent is so entered, whilst the sire or dam of the fourth grandparent must be entered.

21. When one parent is an imported goat entered in the Herd Book, and the sire or dam of the other parent was either a dagger male or a Star or Q Star female goat.

22. When two of the grandparents are entered in the Herd Book—the third is a Star or Q star female goat, whilst the sire or dam of the fourth must be entered.

In 1905 an important departure took place in the introduction of a Toggenburg section. Entries here are not dependent on the Prize Record, but are based on purity of breed, any goat being eligible that is imported from Switzerland or bred direct from imported stock. More recently as other breeds were introduced, further sections had to be incorporated, and there are now separate sections for "Saanens" "Nubians," "Anglo-Nubians," "British Alpines," "British Saanens," "British Toggenburgs," and "English."

Of late years there has been provided a Foundation Book, eligibility for entry in which is governed by the following regulations:

When the goat has won a prize at a Recognised Show in an Adult Class open to general competition, or in a Milking Competition, or has won a Star or Q Star award in accordance with Regulation 83.

When one parent is entered in the Herd Book, and the sire or dam of the other parent is also entered.

When one parent, who must be either a dagger male or a Star or Q Star female goat, is entered in the Herd Book.

When one parent is a Star or Q Star goat not entered in the Herd Book, and the sire or dam of the other parent is entered in the Herd Book.

The progeny of goats entered in the Foundation Book—if sired by, or out of, a goat entered in the Herd Book, whose parents are also entered therein, will be eligible for entry in the Herd Book.

In the case of an imported goat entered in the Herd Book

the condition that both parents should also be entered therein will not apply.

The Kid Register which was established in the very early years of the Society has since been abolished and instead a "Show Register" has been introduced solely as a means of identification for kids, goatlings or goats which are not eligible for entry in the Herd Book proper or the Foundation Book.

The next important step was the establishment of a Register of Stud Goats worthy to be recommended to breed superior stock. Such animals were in those days selected mostly from the Prize Record though a few were passed by inspection these latter being generally imported stock. Later they were chosen mainly through being entered in the Herd Book preference being given to the progeny of "Star" milkers.

The Stud Goat Scheme

The "Stud Goat Register," which has been mentioned in the earlier Editions of this work, was discontinued in its original form after 1922, but a list of Stud Goats still appears annually under what is known as the "Stud Goat Scheme," each animal being personally inspected before acceptance. The "Scheme" is administered by the B.G.S., but is under the approval of, with financial assistance from, the Ministry of Agriculture and Fisheries. Services under this new arrangement are limited, as far as possible, to goats owned by cottagers, smallholders and other persons of a similar position who could not reasonably be expected to pay a fee in excess of 5s., which is the most that is allowed to be charged.

The important features of this new Scheme are: firstly, that applicants for the registration of Stud Goats are not restricted to members of the British Goat Society and secondly, that goats approved for service entitle their owners, under certain conditions, to receive a premium of 5s. for each service including those, up to three in number in one season, of the owner's own she-goats.

The latest date for receiving applications for registration of Stud Goats is 1st June. The season opens on 1st September and closes on the last day of February. If a service proves ineffectual, a second may be claimed, which must be given without further payment within six weeks.

CHAPTER XXVI

GOATS AND THE MALTA FEVER

THE "Malta Fever," or more properly Mediterranean Fever—for it is not restricted to that island—is a matter of such vast importance, closely connected as it is with the goat and its produce, and especially having been the subject of careful investigation and research by a Commission in recent years (1904-6), that it is very desirable that some information on the disease and the enquiry should be included in this book. The brief particulars I am able to give are culled from lectures delivered before the Royal College of Physicians of London on 5th, 10th, and 12th March, 1908, by Dr. J. W. H. Eyre, Bacteriologist to Guy's Hospital, and a member of the advisory board of the Mediterranean Fever Commission, the lectures having been since published in handbook form. I am also able to include some interesting details given to me personally by Dr. Eyre, to whom I am in consequence much indebted.

It is scarcely necessary to state that the Commission was appointed on account of the numerous cases of this peculiar fever that had raged for many years among the troops stationed on the island, and which it is said has been responsible not only for many deaths, but "for a total loss to the combined Army and Navy on the Mediterranean station of at least 75,000 days' sickness per annum."

It is not within my province to describe the clinical symptoms or other technical details of this febrile complaint, but I may state briefly that, as regards the human subject, it is an intermittent, or mostly intermittent, fever resembling enteric. The patient may be comparatively well in the morning, and in the evening show a temperature of 103 or 104 degrees, whilst the next morning his condition may be again normal, the disease extending over long periods, such as three or four months. As to the goats, one of the most striking observations recorded was the healthy appearance of the majority of those infected. We read that "the animals were sleek and plump, with smooth, healthy-looking coats; they took their food well, were as active as their uninfected fellows and yielded as large a quantity of milk, and of apparently as good a quality. . .. In many instances the infected milch goats were the best-looking and the best milkers in the herd, and in a few instances only it was noted that an infected animal suffered from a short, barking cough at infrequent intervals." This disease is, it seems, no new discovery, having been known and studied from very early times, and having been described by Hippocrates. It is due to a microbe known as *Micrococcus melitensis*, which has been shown by various observers to be present in vast numbers in the blood, and as a consequence, in the milk of the goat.

Origin and Distribution of the Disease

Dr. Eyre is of opinion that "*Melitensis septicæmia* [Mediterranean Fever] is primarily a disease of the goat which had its origin in the Persian hills [the Persian wild goat, *Capra ægagrus*], and which accompanied that goat on its world-wide wanderings, remaining potentially active for man so long as its host preserved its original habits in barren rocky countries in the tropics and sub-tropics, where pasturage is of the scantiest, and consists chiefly of shrubs and weeds. When, however, the goat reaches those temperate climes and abundant pasturage which are so pre-eminently suitable to the cultivation of the cow that its lactiferous capabilities fail to rival those of the last-named animal, and selective inbreeding is in consequence neglected, the micrococcus no longer finds a suitable habitat in the caprine mammary glands, and rapidly disappears."

In this connection it may be well to observe that one writer, in discussing the geographical distribution of the fever, remarked upon the fact that "north of latitude 46 degrees N., which corresponds roughly with isotherm 55 degrees F., this fever does not exist as imported cases." Within this limit, however, cases are recorded in various localities far distant from the Mediterranean, the chief seat of the disease. Thus it has been proved to exist as an endemic not only in parts of Europe, such as Trieste, Athens, and the whole of Southern Italy, but in Arabia, Asia Minor, Hong Kong, and many places in India, besides Algeria, Tunis, Egypt, Cape Colony, and South America. The disease disappeared from Gibraltar only a few years ago.

Another member of the Commission is reported to have stated that *Melitensis septicæmia* only occurred in localities in direct sea communication with Malta, implying thereby that the disease was the consequence of the importation of Maltese goats. As an instance of this, a case is recorded of "a herd of sixty-one milch goats (all healthy in appearance and good milkers, many being prize animals) and four male goats which were shipped to the United States in September, 1905, having been purchased by the late Mr. Thompson, of the United States Bureau of Animal Industry. Subsequent bacteriological examination resulted in the recovery of *M. melitensis* first from the milk of two of the goats, and afterwards from that of several more." At intervals varying from eighteen to twenty-four days from the embarkation of the goats eight of the crew of the vessel carrying this cargo were taken ill, and investigation of the state of the blood showed that Mediterranean Fever was the cause. Mr. Thompson himself, who was also on board the same vessel, died some four months later. His death, however, was not ascribed to the fever, but to "bilateral pneumonia, following influenza."

Seasonal Prevalence

Dr. Eyre observes, as an interesting point, that "a milch goat whose serum gave such a pronounced reaction as to make it absolutely certain that infection with *Micrococcus melitensis* had taken place, failed to excrete the coccus in its milk, which by the way, was drying up. This animal was impregnated, and on dropping its kid again yielded large quantities of milk, and now the micrococcus appeared in the milk in large numbers, and was isolated therefrom with the greatest ease. This observation has since been repeated, thus suggesting that the milk of infected animals is most dangerous during the period following the birth of the kid—in other words, at the beginning of summer. . . . At the same time it must be noted that the Maltese goat remains in milk for a very long period, often one or two years, and sometimes as long as three years, and this means that a supply of infective material is always to hand, and that cases should occur throughout the entire year."

Transmission of the Disease

The means of transmission of the disease from goat to goat was carefully investigated, and the experimental work showed that animals could be infected with a fair amount of ease by the ordinary laboratory methods of inoculation, that is, "by the subcutaneous and intravenous injection of living cultivations of *M. melitensis*, and also by the introduction into the system of food artificially or naturally infected." On the other hand, "no experimental results were available with regard to infection by contact." Contagion from one goat to another was shown to have been carried in great measure by means of the goatherds' hands when soiled with infective milk, it being introduced through the skin wherever the least abrasion appeared. The disease can be conveyed in this manner also to the human subject, as well as through the alimentary canal. The Maltese goat has a very long udder, and it often happens when in full milk that the teats actually touch the ground, so that in walking about abrasion of these exposed parts may easily occur if such parts come in contact with excreta or other dirt, thus providing another means of infection. Sexual intercourse between goats is also regarded as responsible for the transmission of the disease, strong evidence in proof of this having been adduced in the course of the observations. Malta Fever was found to be much more prevalent amongst the officers than the rank and file of the Army, and similarly amongst the richer inhabitants of the island than the poor, for the simple reason that "the poor man rarely uses milk as an article of diet, and this is particularly the case with the goatherds themselves, who keep their goats for profit and not to provide food directly for themselves and their families."

FEVER UNKNOWN IN ENGLAND

As the circumstances in connection with this disease may lead people who are not thoroughly acquainted with all the facts to give goats and goats' milk a wide berth, in the fear that this disorder might be introduced into their own households, it is well that I should state, on the authority of Dr. Eyre, that no case is on record of any person having been attacked by this disease from drinking the milk of goats in England. In every instance where symptoms of the fever have been manifested in this country it has been traceable to recent sojourn in Malta or the neighbourhood. Some twenty years or so ago Maltese goats were frequently brought into this country, being landed from yachts and vessels on which the milk from these animals had been used for passengers, and I myself have had several at different times in my own herd; but I never knew of a case of illness arising therefrom. It is quite possible, nay probable, that the microbe to which the disease is due was present in the blood of some of these imported Maltese goats; but as stated above, the coccus can only develop and become a source of danger under climatic conditions which do not obtain in this country.

CHAPTER XXVII

DISEASES, PARASITES, ETC.

AFTER an experience with goats of various breeds extending over a considerable number of years, I have been forced to the conclusion that these animals, under the conditions in which they are usually maintained in this country, are not the hardy creatures they are popularly supposed to be, and which I myself at one time thought them. No doubt in a wild or semi-domesticated state on the rocks and mountains, where they love to roam and where they obtain the kind of food best suited to their requirements, goats, like most other animals in similar circumstances, rarely suffer from disease. But when housed, and especially in badly ventilated stables, or tethered on rich pasturage where the soil is heavy, this hardihood no longer exists, and they become subject to some of the diseases common to sheep and cattle, besides some peculiar to themselves.

When I first wrote on goats I had always been most successful as regards their health, never losing a single one. Up to that date I had never kept more than two or three at a time, and those were to a great extent stall-fed. Subsequently, however, I moved to the neighbourhood of a large common, and with the facilities for pasturage thus afforded, increased my herd to a dozen, letting them roam freely over the ground during the greater part of the year. I then learnt in a very practical manner what I have stated above, and the lesson has been repeated since in other, though similar, circumstances. This opinion, however, is not based on my experience alone, but on that of numerous goat-keepers who, like myself have kept their animals during most of the year on grass. The nature of the soil, as before stated, has a great deal to do with this; a cold clay, insufficiently drained, always operates disadvantageously to these animals sooner or later.

General Indications of Ill-health

The following are some very general symptoms of constitutional disturbance, showing a goat to be out of health: Suspended rumination, or what is commonly called "loss of cud," the nose dry, the horns hot, a swollen face, a staring coat, a tight skin, and an unnatural condition of the fæces. The owner of a sick goat should endeavour to locate the disease, and ascertain what organs or structures are implicated, and to what extent. It is well then to seek for the cause of such symptoms, as if this be removed the effect will generally cease. Medicines should only be had recourse to when every attempt to assist Nature in the work of restoration has failed.

A Wasting Disease

There is a disease which carries off annually a very large number of goats, but which has never yet as far as I am aware been officially diagnosed. It used to be called in early days "A disease peculiar to goats," a vague term no longer employed as the symptoms so closely resemble a disorder in cattle known as "Johne's disease (referred to below) that it is probably the same complaint.

Symptoms.—The first thing often noticed is a falling off of appetite, which may at first be slight, but soon gets worse, until it is a difficult matter to get the animal to eat at all. The result of course, is that it rapidly loses flesh and falls away to a skeleton. Sometimes, however, the appetite remains tolerably good, but the emaciation goes on just the same, though the process is slower. The breathing is sometimes laboured, and the breath nearly always very unpleasant. A cough is very often an accompaniment of the disease, leading one to suppose that, with other symptoms, the lungs are affected, but this is seldom the case. The one prevailing feature which can never be overlooked is the general bloodlessness of the animal. This is shown by the pale colour of the gums and inner surface of the lips, and also of the membranes lining the eyelids, which in health are of a bright red.

Before the disease is far advanced diarrhœa generally

sets in. At first this is not particularly severe; it soon gets worse, however, and though it may be temporarily arrested it is almost sure to break out again, owing to the failure of the digestive organs from the impoverished condition of the system. The diarrhœa may last for several weeks, but as it progresses it often takes the form of dysentery. The poor creature then becomes too weak to stand, and it generally dies uttering plaintive cries and moans. There are other symptoms which occasionally manifest themselves, such as a swelling about the jaws, due to weakness also. The animal always stands with its head down and back arched, looking the picture of misery and dejection.

Strange to say, in many of the post mortems which have been made and reported to me there have been no signs of organic disease. Internal parasites are sometimes discovered, but not in sufficient quantities to affect greatly the animal's health, much less cause its death, which in all cases is due to exhaustion. Both Professor Axe and Mr. A. J. Sewell, who were at one time consulting veterinary surgeons to the British Goat Society, have made numerous post-mortem examinations of goats that have died of this disease, besides treating several in various stages of the disorder, and both have arrived at the conclusion that it must be due to blood-poisoning, brought on by foul herbage, the result of over-pasturing—that is to say, keeping too many goats at one time on one piece of ground, or a few continually on the same plot.

The cases that have occurred in my own herds—and they have been numerous enough—have pointed to this cause; indeed, I could arrive at no other conclusion, for, whenever I have changed my residence, and, consequently, the pasturage, the goats have done well for the first twelve or eighteen months, after which the disorder has reappeared. In one instance, when I rented an extra meadow for a time, the goats in that meadow were doing well, whilst those on the field attached to the house were sickening. Then, again, almost invariably the reports of cases have come from persons keeping goats on pasture, rarely or never from those living in towns, whose goats were entirely stall-fed. Goats in young that are attacked by this disease invariably slip their kids prematurely, and, this further exhausting them, they rarely recover.

TREATMENT.—With regard to treatment, Mr. A. J. Sewell, in a paper on this subject read at a meeting of the British Goat Society some years ago, recommended as follows: "Having first warmly housed the patient, which is very necessary, a mild dose of purgative medicine should be given. I say a mild dose, because the bowels, being in a relaxed condition, are more easily operated upon than if constipation were present. This medicine is given with the idea of removing any irritating matter that may be present in the intestines. Linseed oil and Epsom salts—it does not matter which is given—will do. The dose of oil for an adult goat of moderate size is from 1½ oz. to 2oz.; the dose of Epsom salts is about 1oz. This is best given in a little warm oatmeal gruel, the oatmeal assisting the action of the medicine. After the purgative medicine has had time to operate, astringents must be given. The following is what I have used with the best success, namely, powdered catechu, chalk, and gum, of each ½dr.; powdered ginger, 1sc.; and powdered opium, 6gr. This is one dose, and should be given in a little wheaten-flour gruel, and be repeated two or three times a day, according to the severity of the diarrhœa. As the diarrhœa lessens, the doses should be gradually decreased. It is a bad plan to discontinue the medicine directly the diarrhœa appears to have stopped, for a recurrence of it is often the result. The dose for kids (of the purgative medicine as well as the astringent) is from one-third to half of that recommended for adult goats. When the diarrhœa is obstinate I have given bark with very good results. I have also given this medicine, if the appetite is very bad, after the diarrhœa mixture has been discontinued, as it acts as an astringent as well as a good tonic. The dose is about 30gr. of the powdered yellow cinchona bark. This medicine is also best given in gruel. I have heard that port wine and cayenne pepper are a capital mixture for diarrhœa, and though I have never tried it, I should think it would act well in mild cases. Nitrate of silver I have used with very good results in bad cases when the motions are accompanied with blood, constituting dysentery, the dose being a grain and a half to two grains twice a day. It is best given as a pill or bolus made with bread-crumbs."

When the diarrhœa stage is not arrived at, Professor Axe

relies entirely on iron to strengthen the system and give tone to the blood. He prescribes sulphate of iron in doses of 5gr. each twice a day in water. I have added to this a small quantity of sulphate of quinine when the appetite has been very bad, and found it of great service. The best plan is to procure three or four 8oz. bottles, and put 2sc. of the iron crystals and ½sc. of the quinine into each bottle, giving 1 fluid oz. of the solution—which must be shaken occasionally—for a dose.

FEEDING.—Feeding a goat with this disease is difficult; like ourselves in illness, it will often eagerly devour the very articles that it should not have, and refuse what is good for it. It will rarely, however, eat much of anything. All green fodder should, of course, be withheld when diarrhœa is present, though it will generally be taken readily when offered, probably for the sake of the change. Hay must constitute its chief food, and although clover will often be preferred, good meadow hay I consider best, as the woody fibre in the sticks of clover must of necessity be more indigestible. I have had goats with this disease which for a long time subsisted on nothing but hay and bread and milk, taking the latter most readily; they would each drink a quart or more of milk in a day. When the diarrhœa was present I mixed arrowroot or Oswego cornflour with the milk. Goats will often take sloppy food like this very readily in illness when they would not touch it in health. Dry bread is usually eaten with a relish at this time, though corn of any kind is nearly always refused.

Mr. Arthur Gill, M.R.C.V.S., of Bexley Heath, one of the late veterinary surgeons to the British Goat Society, and one who had had much practical experience with goats, both in health and disease, finds that when taken early the best treatment for the disorder under discussion is the following: Put half-a-pint of rice into two quarts of water, and boil down to one and a half pints of liquor, strain away the grain, and when the liquor is cool enough, and getting fairly thick, take half a pint, and add sulphurous acid forty minims and tincture of opium one drachm, giving this thrice daily. Feed on nourishing diet, which must be at the same time easy of digestion. It may be advisable to mention here that the British Goat Society is taking further measures to investigate this insidious disease.

Jöhne's Disease

Jöhne's disease, which has no doubt been long confounded with Takosis in this country, is a disorder of a wasting character which was first noticed in cattle, though it seems to have more than one form. It has been found to be due to "an acidfast bacillus possessing the same staining reactions as the tubercle bacillus and indistinguishable from it in size and shape." The successful cultivation of this bacillus was achieved about the year 1910 at the Brown Institute, when a careful research threw much light on the subject and revealed, amongst other things, the fact that goats are liable to it.

A writer in *Veterinary Notes* says: "The course of Johne's disease is very slow, and the first signs are often unrecognised. Well-fed animals lose flesh and do badly and yet show no signs of fever, cough, or loss of appetite. Later they suffer from periodical attacks of diarrhœa, which increase in severity and frequency until the animal becomes terribly emaciated and dies of starvation." It is useful to know, as I gather from these notes and other sources, that as a result of the cultivation of the bacillus a diagnostic vaccine has been prepared which should be sufficiently specific for practical purposes, and in fact as specific as tuberculin is for tuberculosis. The culture is prepared at the Royal Veterinary College, and has been termed "Johnin." Experiments with goats have shown not only the susceptibility of these animals to Johne's disease, but the efficacy of the vaccine when sub-cutaneous inoculation has been performed. This disease is communicable from an infected animal to a healthy one, and owners should separate from the herd any goat manifesting symptoms of this malady, and at once disinfect the place in which they are kept. See also "Takosis."

Catarrh, or Influenza

Goats that are accustomed to being kept constantly in warm stables sometimes suffer, if much exposed to damp and cold, from disease of the bronchial tubes and affections of the respiratory organs. The symptoms of catarrh are copious discharges from the nose and eyes, accompanied by sneezing; the eyes have frequently a bloodshot appearance. With bronchitis there is considerable difficulty in breathing, and a wheezing cough is present. The treatment consists in keeping the animal warm and quiet, feeding it on gruel and mashes, which, with a few doses of Epsom salts and ginger, will generally effect a cure. The proportions for one dose are about 1oz. of Epsom salts to 1 drachm of ginger.

Constipation

Constipation of the bowels only appears to any extent in kids, and, like diarrhœa, occurs mostly after weaning, when the stomach receives a change of food. It is not often dangerous, however, and will generally yield to a mild aperient in the form of an ounce of common salt, or to one slightly stronger, as ½oz. Epsom salts, administered in a quarter of a pint of either warm water or gruel. The diet should be of a relaxing character, with a bran mash every other day. Particular care should be taken to have plenty of rock salt accessible.

Diarrhœa and Dysentery

The outward appearances of these disorders being very similar, I class them together, although in reality they differ materially, the former meaning simple purging, and being a natural effort of the bowels to get rid of irritating matter; whereas the latter consists in inflammation of the mucous membrane of the intestines, causing violent purging, and accompanied generally by fever and bloody evacuations. As dysentery frequently results from neglected diarrhœa, it is well not to allow the latter to continue long before adopting remedial measures. It often happens, however, that a cure is effected, if taken in hand early, by simple change of food, or even of pasture; thus I have known cases of purging caused by giving too much corn, and have cured them with green meat without any medicine. Should this have no effect, then medicine must be administered. I have found this disorder very prevalent with kids that are brought up by hand, it being caused by giving the animal too great a quantity of milk after it has fasted.

The treatment I have pursued with success in such cases consists of an aperient combined with a gentle tonic. This may be made by mixing together ½oz. of Epsom salts and

½ drachm of ginger, to which is added one tablespoonful of sheep's cordial, consisting of equal parts of brandy and sweet spirits of nitre. This is to be given when the evacuations are of a yellowish-white appearance. When, however, they are dark green, or, still worse, black, in which case the matter is more serious, give an ounce of castor oil with a teaspoonful of oil of turpentine. Should the diarrhœa be but slight, and the creature lively, with its usual appetite, no notice need be taken of it; but when the reverse is the case, then physic should be given without delay.

Diarrhœa in full-grown goats is very common in spring, especially with those that have been stall-fed all the winter, and kept chiefly upon dry food. The fresh spring grass in March and April is certain to cause scouring unless introduced very gradually, and in very small quantities at a time; and it is often some days before the bowels can be got into their normal condition again. This change, in fact, should not be brought about suddenly, but by degrees, it being frequently necessary to clear the bowels with a dose of linseed oil. For full-grown goats, if the purging be considerable, give ¼oz. of prepared chalk, with two grains of opium, in a pint of warm milk, and repeat the dose in two days if the disorder continues.

There are several remedies besides chalk for this complaint; sulphate of iron, for instance, is useful, given in small quantities, especially if debility is present. Baked flour is also a good thing to give in such cases, with rice-water as drink.

In dysentery the discharges are thin and slimy, being frequently mixed with blood and hard lumps, and very offensive; the creature becomes weak and emaciated, and generally refuses food. This disease may be brought on by the goat eating decayed and decomposed vegetables, or by irregular feeding, or may follow neglected diarrhœa. The animal must be kept warm in the stable, and have a good bed of straw and plenty of hay. If it refuse the hay, as it probably will if very ill, gruel must be administered to it. In the way of physic, ½dr. of rhubarb with 1oz. of Epsom salts, or 2oz. of linseed oil with 2gr. of opium in half-a-pint of linseed tea, must be given to clear the bowels, and then the following mixture to act as an astringent: Prepared chalk, 1oz.; powdered catechu, ½oz.; ditto ginger, 2dr.; ditto opium, ½dr.; peppermint water, half-a-pint. Mix these ingredients well together, and give from one to two teaspoonfuls twice a day, shaking the bottle previously. Afterwards a tonic composed of powdered gentian and ginger, commencing with ½dr. of the former and 1sc. of the latter, and terminating with four times these proportions, given in water, is valuable to restore and strengthen the system.

Foot-Rot

When kept on low, wet ground, goats are liable, like sheep, but less frequently, to attacks of this disease, in which the hoof outgrows its natural proportions, not having that friction which on rocks and hard ground wears down the horn as it forms. The outer portions of the hoof outgrowing the inner parts, or frog, the former extend over the latter, which becomes soft, cracks, and lets in dirt and sand; these penetrate to the quick, causing great irritation and often ulceration. Inflammation having set in, the coronet swells, and portions or sometimes even the whole of the hoof become disorganised and break away.

If taken in time this disease may be treated successfully, but the task is a troublesome one and unpleasant to perform. The hoof must be well and carefully pared with a sharp knife, taking care, as far as possible, not to damage the quick or bed of the horn. The part should then be thoroughly cleansed from all dirt with warm water, and dressed with a solution of carbolic acid consisting of one part of the acid to twenty parts of glycerine. A poultice of linseed meal may then be applied for one or two days, after which the foot should be enveloped in cotton wool or tow, and enclosed in a coarse bag. The carbolic dressing is to be repeated once daily. If, in the course of healing, proud flesh (i.e. red wart-like excrescences) should appear, this must be overcome by caustic dressings, of which nitrate of silver is the most effectual. Application of the caustic may be made once daily until the excess of growth has been overcome and the surface of the wound brought to its proper level. Some recommend medicine to be given internally at the commencement of the treatment, and administer Epsom salts 2oz., sulphur 1oz. Foot-rot is a disease, however, which may be prevented by occasionally examining the hoofs and paring away any horn that has exceeded its ordinary proportion.

GARGET.

Garget, black garget, or inflammation of the udder, is a serious ill, and if not taken in hand early and properly treated, is very likely to end fatally. It may be brought on owing to the goat getting chilled by lying on damp, cold ground after kidding, by blows to the udder, by bad milking, or as the result of sore teats that have been neglected. The first symptom of this disease is the refusal on the part of the mother to allow her young to suck. When this is noticed the udder must be examined, and if the part feels hot and hard, it is a further sign. The goat must be separated from the kid and kept shut up in a warm, well-sheltered place. The udder must be emptied frequently. A dose of Epsom salts should be administered, and repeated in three days, the udder, in the meantime, being freely bathed with hot salt-and-water. When the inflammation and swelling have subsided, the gland may be gently rubbed with soap liniment once daily. If this does not arrest the disorder, but the udder turns black, the services of a veterinary surgeon must be obtained.

LICE

These are disgusting and troublesome pests, and, with goats that are much petted by children, especially objectionable. They often appear in large numbers in the hair of goats or kids that have got into a low and poor condition from insufficient food or sickness. A good and simple remedy is to sponge the animal thoroughly with tobacco-water to which a little spirits of tar has been added, the proportion being a wineglassful of the latter to a quart of the decoction of tobacco. Mercurial ointment one part, mixed with seven parts of lard, is a certain killer of these vermin, but it is a nasty process and a dangerous one, the ointment being very poisonous. An indiarubber syringe charged with some insect-destroying powder, such as that sold by Keating, which on pressure is blown into the hair of the animal, has been employed with advantage in some cases. If goats are properly fed and cared for, however, and their coats regularly brushed and combed out, such vermin will not establish themselves.

Two other remedies have been found recently to be very effective. One of these is powdered stavesacre (seeds of the Russian larkspur) boiled in water containing a little washing soda. The proportions are 4oz. of powdered stavesacre, 2oz. of washing soda, and 1gal. of water boiled for an hour. The dressing should be applied to every part of the body from the tip of the nose to the tip of the tail. It kills the live insects but another dressing should follow in four or five days to destroy those hatched from the nits since the first application. This is said to be better and safer than paraffin or sheep-dip. At the same time it must be remembered that stavesacre is a powerful poison. There is also a veterinary preparation called Bacterol, supplied by the Bacterol Company, Ltd., of Highgate, N., which has been recommended to me as very efficient as a destroyer of these pests.

Rheumatism

This is an ailment in goats which is far from uncommon but I have never been able to find a remedy for it that could be regarded as a cure. A writer, however, in *The Goat World*, a Californian journal devoted entirely to goats and goat-keeping, mentions a case of rheumatism in a kid which was successfully treated by giving "small doses of the usual family remedy for this complaint, viz., sodium salicylate tablets, made with wintergreen oil." One tablet was cut into four parts, dissolved and given in milk, and the four doses were administered the same day. At the end of two days an improvement was noticed, and in a week the kid was running about quite cured.

Foot-and-Mouth Disease

This disease is often referred to in connection with importation of goats, the fear of its possible introduction into this country being frequently mentioned as the reason of the Ministry of Agriculture refusing to grant facilities for such importation.

It cannot be asserted that goats are immune from foot-and-mouth, because cases are known to have occurred in Switzerland and other parts of Europe, but I have never heard of a single case in the British Isles.

Poisons

Although goats will consume with impunity many herbs which, if eaten by other animals, would rapidly cause death, there are, nevertheless, a few which are equally injurious to them. Amongst these are the leaves of the yew, which, if taken in any quantity, will, especially when the branches are dead, cause violent sickness. If treated at once, the poison may be got rid of by administering strong doses of purgative medicine, such as linseed or castor oil, but a more effectual remedy, especially in bad cases, is the stomach pump.

Sore Teats

The teats of a goat are sometimes attacked with sores and ulcers, rendering both milking and the natural sucking of the kid most painful operations to her. The soreness is generally produced by the kid biting the teat, but other, and probably constitutional, causes may bring it on. Although it may be considered a comparatively unimportant affair, it nevertheless requires attention, or, like other simple disorders, it may, if neglected, lapse into a more serious evil. The kid not being able to suck, the milk accumulates in the udder, and, as remarked above, garget may ensue. The young goat must be temporarily removed from its mother, and fed by hand until the latter is able to suckle it. The teats should be dressed with the following ointment: Goose grease, 8oz.; camphorated spirits of wine, 2oz.; common salt, 2oz. In mild cases, where they are but slightly cracked, goose grease alone will often effect a cure. I have latterly, however, used vaseline for this purpose, and find it answers perfectly. The milk must be drawn off twice daily, but the operation should be performed as gently as possible.

Takosis

A pamphlet has recently come into my hands which has a most important bearing on the disease peculiar to goats referred to previously. It treats of "Takosis, a Contagious Disease of Goats: Being a Preliminary Report on its Nature, Cause, and Prevention, by John R. Mohler, V.M.D., A.M., Chief of the Pathological Division, Bureau of Animal Industry, and Henry J. Washburn, D.V.S., Acting Assistant Chief." The work is issued officially by the U.S. Department of Agriculture at Washington. It is evident from the contents of the forty-four pages comprising this pamphlet that what I referred to above as "a peculiar disease," and "takosis," the name given in America to a disease which appears to have originated at a farm of Angora goats in Pennsylvania, are practically one and the same. The name is indeed well chosen, being derived from a Greek word meaning "to waste, to cause to waste away." A history of the outbreak, together with the symptoms of the disease, its treatment, and other matters of interest, appeared in *The Bazaar, Exchange and Mart.*[1]

No specific appears to have been found for this disorder, but under the heading "Therapeutics" we read as follows: "The most pleasing results that have been derived from the use of drugs in our experiments at the laboratory have followed the administration of calomel given alone in 0.10 gram doses twice daily for two days, to be followed by powders composed of arsenic, iron, and quinine, as follows:

Arsenious acid	1.40 grams.
Iron (reduced)	12.00 ,,
Quinine sulphate	6.00 ,,

Mix and make into twenty powders, giving one to each adult goat morning and evening at the conclusion of the administration of calomel. After an interval of two days this treatment is repeated. In case the diarrhœa persists, the sulphate of iron has been substituted for the reduced iron with beneficial effects."

[1] These articles have been reprinted, and are among the "Leaflets" issued by the British Goat Society.

The following are the conclusions arrived at as the result of the "present preliminary investigation":

"1. The disease here described as takosis has appeared in many parts of this country, but particularly in the Northern States, where it has caused great loss to many breeders of Angora goats.

"2. It is a progressive, debilitative, contagious disease, characterised by great emaciation and weakness, with symptoms of diarrhœa and pneumonia, and causes a mortality of 100 per cent. of those affected, and from 30 to 85 per cent. of the whole flock.

"3 From the carcases of numerous animals that have succumbed to this disease a new organism, *Micrococcus caprinus*, has been recovered in purity, and is presumably the etiological factor.

"4. This micrococcus possesses pathogenic properties for goats, chickens, rabbits, guinea pigs, and white mice, but not for sheep, dogs, or rats.

"5. Although the disease has been described before [here a reference is given to "The Book of the Goat" (third edition), 1885, p. 212], so far as could be ascertained no bacteriological investigations have been previously made.

"6. Medical treatment was attempted with varying success, while the immunising experiments thus far conducted (although too few to permit of any conclusive statement or accurate estimate as to their protective value) have shown highly encouraging results. When accompanied with measures of isolation and disinfection the treatment may prove of great assistance in the suppression and eradication of the disease in an infected flock."

Administering Medicines

A few words of advice on this subject may be useful, as success depends upon the manner in which medicines are administered. Those of a fluid consistence (and they should nearly always be administered in this form) should be made to trickle down the gullet as slowly and gently as possible, for, if poured hastily into the animal's throat, instead of remaining in the fourth stomach or abomasum, as it should do, it passes into the rumen, where it lies inert, and is therefore comparatively useless. A drench-horn is the proper utensil to employ in administering fluids, but an old teapot with a small spout answers the purpose very effectively. I have had goats which would drink linseed tea or gruel like water, but such accommodating animals are not common.

CHAPTER XXVIII

DISBUDDING—PREVENTING GROWTH OF HORNS

Dishorning

Accidental dishorning occurs occasionally, and is very alarming. I have had two cases in my experience. In one of these the goat got its horns caught in the iron bars of a hay rack, and in its attempts to free itself gave a sudden wrench, which detached the shell, or case, from the core. This core, when thus exposed, had very much the appearance of the fleshy, or edible, part of the claw of a boiled lobster. It was a bright red, being covered with blood, which trickled rapidly down the face of the goat, presenting a horrible appearance. I was recommended by a neighbouring farmer, to whom I applied for advice on the matter, to smear the horn with tar, he meaning *Stockholm* tar; but as this was not specified, and there happened to be some gas tar on the premises, it was used instead. Never shall I forget the cries of pain of that poor goat, whose sufferings I was quite unable to alleviate, for, once applied to the raw surface, it was impossible to remove the tar. However, this effectually cauterised the wounded part, and in course of time the core resembled the other horn. No great difference could be detected, except that one horn was somewhat shorter and of a rather different shape. I have since understood that the application of Stockholm tar, although, of course, a painful process, does not cause the same amount of suffering, and is the best remedy to apply. A strand of jute or tow is saturated with this tar, and is bound round the core.

As to intentional dishorning, it is a practice to be strongly condemned as cruel. I have seen it done in one case, and the pain caused to the goat was such that I should never allow the operation to be performed on a goat of my own.

Preventing the Growth of Horns

This is quite a different matter from dishorning, and can be carried out with very little trouble, whilst causing but slight pain, if any, to the kid, for it must be performed when the animal is quite young. As will be seen, the directions here given were intended to be applied to calves, but they are equally applicable to kids. At the same time it must be stated that the effect is not so certain in the caprine as in the bovine species. These directions are extracted from a leaflet published by the Ministry of Agriculture:

A Substitute for Dishorning.—"Clip the hair from the top of the horn when the calf [or kid] is from two to five days old. Slightly moisten the end of a stick of caustic potash with water or saliva (or moisten the top of the horn-bud), and rub the tip of each horn firmly with the potash for about a quarter of a minute, or until a slight impression has been made on the centre of the horn. The horns should be treated in this way from two to four times, at intervals of five minutes. If, during the interval of five minutes after one or more applications, a little blood appears in the centre of the horn, it will then be only necessary to give another very slight rubbing with the potash.

"The following directions should be carefully observed: The operation is best performed when the calf [or kid] is under five days old, and should not be attempted after the ninth day. One man should hold the animal whilst an assistant uses the caustic. Roll a piece of tinfoil or brown paper round the end of the stick of caustic potash, which is held by the fingers, so as not to injure the hand of the operator. Do not moisten the stick too much, or the caustic may spread to the skin around the horn and destroy the flesh. For the same reason keep the animal from getting wet for some days after the operation. Be careful to rub on the *centre* of the horn, and not round the side of it.

"*Note.*—Caustic potash is *poisonous*, and must therefore be kept in a safe place. It can be obtained from any chemist in the form of a white stick, and when not in use should be kept in a stoppered glass bottle in a dry place, as it rapidly deteriorates when exposed to the air."

Disbudding a Defect

The whole question of disbudding has been the subject of much controversy by members of the British Goat Society, some being violently opposed not only to dishorning, which —being illegal—was generally condemned, but to the preventive process mentioned. This objection was taken on the ground of cruelty, an opinion formed mainly by persons who had seen the effect of the operation when done by careless servants, or by those who had been told of such effects.

As regards the British Goat Society, an attempt was made to render a disbudded goat ineligible for competition at shows, but the decision of the majority left the matter very much to the discretion of the judge. The following is the resolution that now governs this question: "Persons making entries in the Herd Book Kid Register or Stud Goat Register shall distinctly state on the entry form if the goat they propose to register is horned or hornless, or has been dishorned or disbudded. The fact of a goat having been disbudded to be regarded as a defect whenever the animal is being judged."

INDEX

T

U

V

W

Z